Postharvest Technology - Recent Advances, New Perspectives and Applications

Edited by Md Ahiduzzaman

Published in London, United Kingdom

IntechOpen

Supporting open minds since 2005

Postharvest Technology - Recent Advances, New Perspectives and Applications
http://dx.doi.org/10.5772/intechopen.95208
Edited by Md Ahiduzzaman

Contributors
Theophilus M. Maduabuchukwu Ikegwu, Clement C. Ezegbe, Chioke A. Okolo, Chigozie E. Ofoedu, Mahesh Prasad Thakur, Chandra Shekhar Shukla, Harvinder K Singh, Atma Nand Tripathi, P. Sudha, P. Rajkumar, A. Astina Joice, I.P. Sudagar, R. Arulmaru, Kofi Owusu-Akyaw Oduro, Shepard Ndlela, Nelson L. Mwando, Samira A. Mohamed, Trina Adhikary, Ch. Durga Hemanth Kumar, Zhongli Pan, Chang Chen, Ryusuke Oishi, Barkat Hussain, Rayees Ahmad, Shafiya Hassan, Showkat Ahmad, Syed Nighat, Yendrambamb K. Devi, Kounser Javeed, Salma Usmani, Mohd Javid Ansari, Sait Erturk, Mustafa Alkan, Federico Félix Hahn Schlam

First published in London, United Kingdom, 2022 by IntechOpen
IntechOpen is the global imprint of INTECHOPEN LIMITED, registered in England and Wales, registration number: 11086078, 5 Princes Gate Court, London, SW7 2QJ, United Kingdom
Printed in Croatia

British Library Cataloguing-in-Publication Data
A catalogue record for this book is available from the British Library

Additional hard and PDF copies can be obtained from orders@intechopen.com

Postharvest Technology - Recent Advances, New Perspectives and Applications
Edited by Md Ahiduzzaman
p. cm.
Print ISBN 978-1-83969-923-8
Online ISBN 978-1-83969-924-5
eBook (PDF) ISBN 978-1-83969-925-2

For EU product safety concerns:
IN TECH d.o.o., Prolaz Marije Krucifikse Kozulić 3, 51000 Rijeka, Croatia, info@intechopen.com or visit our website at intechopen.com.

Meet the editor

Professor Md. Ahiduzzaman, Ph.D., obtained a bachelor's degree in Agricultural Engineering from Bangladesh Agricultural University in 1993, an MSc in Energy Systems and Management from University of Flensburg, Germany, in 2006, and a Ph.D. in Mechanical Engineering from Islamic University of Technology, Bangladesh, in 2011. He completed his postdoctoral studies at the University of Alberta, Canada. He was awarded a DAAD Scholarship (Germany) and NSERC Postdoctoral Fellowship (Canada). Dr. Ahiduzzaman has been researching and teaching in the field of agricultural engineering, focusing on the processing of field crops, the processing technology of fruits and vegetables, drying technology, renewable energy, and framework development for greenhouse gas mitigation policies in agriculture and other sectors. He has experience working in multi-organizational and interdisciplinary teams, evaluating research, and developing projects in the agricultural engineering sector. He has more than sixty publications in scientific journals and several books and book chapters to his credit. He has had a distinguished academic and research career for more than twenty-four years.

Contents

Preface

Postharvest technology is used to manage crops after harvest to maintain the quality of food. After harvesting, crops lose their self-protection systems, continue metabolism activities, and are prone to microbial attack; as a result, crops begin to spoil. Postharvest management includes processing, preserving, drying, storing, and protecting crops from pests and microorganisms. As such, postharvest management supports food safety and security by reducing losses. There has been much research conducted worldwide on food and safety issues. This book, *Postharvest Technology - Recent Advances, New Perspectives and Applications*, addresses some important aspects of postharvest technologies of food products. It is organized into four sections containing eleven chapters.

The introductory section opens with the chapter "Challenges and Measures to Recapitalise Handling of Postharvest Crops in Developing Countries" by Dr. Ryusuke Oishi. The chapter is an introduction to the constraints and potential measurements of recapitalization of handling and postharvest for food storage, processing, and distribution. The chapter also describes long-term utilization of capital in postharvest handling to reduce losses. Postharvest loss in developing countries is mainly attributable to a lack of capital and technology for food storage, processing (i.e., threshing, drying, and packaging), and distribution. This chapter investigates the causes and potential measurements of postharvest losses in developing countries.

Section 2, "Postharvest Preservation Technology for Field Crops", contains two chapters: Chapter 2, "Postharvest Preservation Technology of Cereals and Legumes" by Theophilus M. Ikegwu, Clement C. Ezegbe, Chioke A. Okolo and Chigozie E. Ofoedu, and Chapter 3, "Stored Grain Pests and Current Advances for Their Management" by Rayees Ahmad, Shafiya Hassan, Showkat Ahmad, Syed Nighat, Yendrambamb K. Devi, Kounser Javeed, Salma Usmani, Mohammad Javed Ansari, Sait Erturk, Mustafa Alkan, and Barkat Hussain. Both chapters address preservation technologies for cereal grains and legumes and describe management techniques for protecting stored grain from pest attack.

Section 3, "Postharvest Disease Management of Fruits and Vegetables," includes Chapter 4, "Robotic Heat Treatments for Mango and Prickly Pear Increase Shelf Life and Reduce Pathogen Infection" by Federico Félix Hahn Schlam; Chapter 5, "Postharvest Diseases of Vegetable Crops and Their Management" by Atma Nand Tripathi, Shailesh Kumar Tiwari, and Tushar Kanti Behera; and Chapter 6, "Advances in Postharvest Disinfestation of Fruits and Vegetables Using Hot Water Treatment Technology-Updates from Africa" by Shepard Ndlela, Nelson L. Mwando, and Samira A. Mohamed. Postharvest diseases threaten food safety, quality, and security. Proper management of postharvest diseases is important for protecting crops. These three chapters identify postharvest diseases and describe remedies for lengthening the shelf life of selected fruits and vegetables.

Finally, Section 4, "Postharvest Processing and Packaging", includes Chapter 7, "Advances in Postharvest Packaging Systems of Fruits and Vegetable" by Trina Adhikary and Durga Hemanth Kumar; Chapter 8, "Postharvest Technology of Tamarind" by P. Sudha, P. Rajkumar, A. Astina Joice, I.P. Sudagar, and R. Arulmari; Chapter 9, "Processing of Tree Nuts" by Chang Chen and Zhongli Pan; Chapter 10, "Edible Coating" by Kofi Owusu-Akyaw Oduro; and Chapter 11, "Postharvest Processing, Value Addition and Marketing of Mushrooms" by Mahesh Prasad Thakur, Harvinder K. Singh, and Chandra Shekhar Shukla. Developing food products from agricultural produce and then packaging and coating the products is important for shelf life and marketing. Chapters in this section discuss the processing technology, packaging, and marketing of selected fruits and vegetables.

Overall, this book presents important information on postharvest technology from internationally reputed authors. It is designed for policy-makers, producers, food processors, industry workers, researchers, and other stakeholders. It is a useful resource for learning about the applications of postharvest technology for reducing losses and enhancing the shelf life of crops. It also serves as a guide to future researchers in the important area of postharvest technology.

Md Ahiduzzaman
Department of Agro-Processing,
Bangabandhu Sheikh Mujibur Rahman Agricultural University,
Gazipur, Bangladesh

Section 1

Introduction

Chapter 1

Challenges and Measures to Recapitalise Handling of Postharvest Crops in Developing Countries

Ryusuke Oishi

Abstract

Global population growth and environmental burdens have caused rising concerns regarding future food security. Contradictorily, many crops are discarded at postharvest stages without being consumed. Postharvest loss in developing countries is mainly attributable to a lack of capital and technology for food storage, processing (i.e. threshing, drying and packaging) and distribution. This study endeavours to investigate the causes and the potential measurements of postharvest losses in developing countries. Specifically, limited budgets in developing countries cannot finance the cost of capital investment; therefore, reliance on third parties such as international organisations is considered a realistic measurement. This investigation establishes that in some cases, a lack of knowledge and skills can result in a lack of full utilisation of the capital provided for handling post-harvest crops. Supporters are discouraged from providing development assistance in circumstances in which whether sufficient results will be achieved is unclear. This study emphasises that enabling the successful long-term utilisation of capital for postharvest handling is critical to improving the rate of vital crop loss.

Keywords: postharvest technology, storage technology, recapitalisation, development support, agricultural investment ratio, development flows to agriculture

1. Introduction

Recent years have brought about increasing opportunities to explore approaches for sustainability around the globe. With the ever-increasing global population and ongoing effects of climate change, addressing future food security has become a major challenge. This is also a major issue taken up by the United Nations. In particular, Sustainable Development Goal (SDG) 2 of the Post-2015 Development Agenda aims to 'end hunger, achieve food security and improved nutrition and promote sustainable agriculture' [1]. SDG 2 emphasises the improvement of food supplies in developing countries, stating, 'Increase investment, including through enhanced international cooperation, in rural infrastructure, agricultural research and extension services, technology development and plant and livestock gene banks

IntechOpen

in order to enhance agricultural productive capacity in developing countries, in particular least developed countries' [1]. Improvement in food supplies tends to be the focus of attention on ensuring food security; however, managing postharvest crops is also a critical issue. This research focuses on postharvest crop handling, analysing the current circumstances and proposing potential measures for improvement.

In many countries, a large number of crops are discarded at the postharvest stage. This occurs because many foods expire before being delivered to consumers. This challenge is considered to be particularly serious in developing countries mainly due to a lack of capital and technology for food storage, processing (i.e. threshing, drying and packaging) and distribution. Improvements in storage and processing technologies can delay spoilage, and efficient supply chain logistics can enable faster delivery of postharvest crops. The achievement of such improvements requires recapitalisation of postharvest technology; however, simply recapitalising postharvest technology will not improve the situation because local people in many developing countries often face difficulties with fully utilising modern postharvest technologies due to a lack of knowledge and capability. Moreover, when successful investment results (i.e. reduction in postharvest losses) are rarely realised, foundation bodies (i.e. international organisations and investors) are discouraged from providing funding in such countries. This analysis seeks to investigate and explain the cause of the circumstances in which the capital for handling postharvest crops is not fully utilised and how this affects foundation bodies' decision-making on providing postharvest development support.

This chapter is organised into six sections. Section 2 presents an analysis of the current circumstances of postharvest loss, and Section 3 investigates the process of handling postharvest crops. In Section 4, the potential for improving the management of postharvest crops through recapitalisation is considered. Section 5 presents a theoretical analysis of capital investment for handling postharvest crops. In this section, we also study three successful cases of agricultural development support in developing countries. Section 6 analyses the agricultural investment and development in the least developed countries and the G20 countries. Finally, Section 7 concludes this study.

2. The current circumstances of food loss at the postharvest stage

This section will analyse the current circumstances in which production is discarded in the postharvest stage. To formally analyse this situation, defining first the ultimate goal of global food security, as well as food loss and waste, is required. The Food and Agriculture Organisation (FAO) of the United Nations indicates that food security occurs when all people—at all times—have physical, social and economic access to sufficient, safe and nutritious food that meets their dietary needs and food preferences for an active and healthy life [2]. Two main issues arise for securing food: improving the crop harvesting process and reducing postharvest loss. The problem with postharvest loss is especially serious, a large number of postharvest crops are discarded without being consumed. One study argues that solutions for reducing postharvest losses require relatively modest investment and can result in higher returns than increasing crop production to meet food demand [3]. Postharvest loss is divided into food loss and food waste. 'Food loss is the decrease in the quantity or quality of food resulting from decisions and actions by food suppliers in the chain, excluding retailers, food service providers and consumers' [4]. 'Food waste refers to the decrease in the quantity or quality of food resulting from decisions and actions by retailers, food service providers and consumers' [4]. In other words, food loss refers to the discarding of food before shipping, and food waste refers to discarding food after its shipping for reasons such as being unsold. The status of postharvest loss differs by country. **Tables 1** and **2** present the amount of cereal product and loss (in 1000 tonnes) in the world's least developed and G20 countries in 2019.[1]

Countries	Maize (product)	Rice (product)	Wheat (product)	Maize (loss)	Rice (loss)	Wheat (loss)
Afghanistan	107	352	3,613	16	25	542
Angola	2,765	10	3	307	0	0
Bangladesh	3,288	54,416	1,099	250	3,104	238
Benin	1,510	459	0	378	115	0
Chad	438	260	2	36	10	0
Ethiopia	8,350	144	0	252	3	168
Madagascar	215	4,030	0	10	403	0
Malawi	2,698	112	6	529	5	0
Mali	3,625	3,168	1	218	127	21
Mauritania	16	232	29	1	7	22
Mozambique	1,250	134	7	75	5	1
Myanmar	1,984	27,574	21	90	861	24
Nepal	2,473	5,152	116	248	486	195
Niger	30	102	1,958	1	4	0
Rwanda	410	120	5	49	3	0
Senegal	264	763	11	33	30	6
Sierra Leone	23	920	0	1	120	0
Sudan	45	3	0	4	2	48
Uganda	2,773	145	0	175	5	20
United Republic of Tanzania	6,273	246	23	738	30	3
Zambia	2,395	43	114	72	2	3

Source: FAOSTAT (http://www.fao.org/faostat/en/#data).

Table 1.
Food products and losses in the world's least developed countries in 2019.

[1] Since there are many types of food, typical crops (i.e. maize, wheat and rice) are used in this study. The least developed countries in **Table 1** are identified by the data source. Countries not reporting data for any at least one of the three crops are omitted. The European Union (EU) is omitted in **Table 2** because the EU does not represent a single nation. The variables are identified in FAOSTAT for Product with the Domain code FSB, Element code of 5511 and Item codes for maize–2514, rice–2807 and wheat–2511. Loss is represented by Domain code FBS, Element code 5123 and Item codes for maize–2514, rice–2807 and wheat–2511. The codes (other than area codes) of variables in **Table 2** are identical to those in **Table 1**. The area and year codes of the variables are not presented because they depend on the countries and years selected. For more information about these codes, please refer the data source.

Countries	Maize (product)	Rice (product)	Wheat (product)	Maize (loss)	Rice (loss)	Wheat (loss)
Argentina	43,462	1,368	18,539	633	53	316
Australia	387	635	20,941	2	6	209
Brazil	82,288	11,749	5,422	8,321	1,181	247
China	257,349	214,079	131,690	11,806	8,624	2,901
France	12,667	73	35,798	112	3	326
India	27,820	172,580	99,700	2,785	4,654	5,987
Italy	6,179	1,512	6,933	11	34	49
Japan	0	9,728	766	4	190	163
Mexico	27,170	284	2,943	4,571	53	197
Republic of Korea	78	5,195	26	205	481	23
Russian Federation	11,419	1,038	72,136	115	21	433
Saudi Arabia	45	0	518	92	0	35
South Africa	12,510	3	1,900	569	0	85
Turkey	5,700	940	20,000	202	30	2,133
United States of America	364,262	10,153	51,398	17,864	398	2,334

Source: FAOSTAT (http://www.fao.org/faostat/en/#data).

Table 2.
Food products and losses in the G20 countries in 2019.

As shown in **Tables 1** and **2**, food losses occur in both the least developed and the G20 countries[2]; however, the causes of food loss differ between them. Specifically, the main cause of food loss in the least developed countries is inadequate equipment and provisions for managing postharvest crops, whereas in G20 nations food loss is likely due to consumers' excessive demands for food quality and retail stores' sales strategies. For example, consumers in Japan demand excessively high food quality (i.e. consumers prefer to purchase attractive produce without pesticides); hence, much of the product that does not suit consumer preferences (i.e. slightly damaged or of an undesirable size) is discarded [5]. There are a large number of extremely competitive retail stores known as convenience stores in Japan that often overstock foods because they do not want to lose customers due to running out of food stock [5]. In contrast, due to insufficient capital for processing and preserving food in developing countries, fresh foods are rarely delivered to consumers. For example, most postharvest grains in developing countries are stored in traditional storage structures that do not prevent insect infestation and mould [3]. Although people try to make the best use of the food produced in developing countries, a significant amount of production is lost in postharvest operations due to a lack of knowledge, inadequate technology and/or poor storage infrastructure [3].

Section 2 analyses the contrasting circumstances of discarding postharvest crops in developed and developing countries. Based on the data investigated, a significant number of cereals are discarded in both developed and developing

[2] According to FAOSTAT, production is reported at the farm level for crop and livestock products, whereas loss represents wastage during the year at all stages at which production is recorded and the household. As a result, in the case of some countries, loss exceeds production.

countries; however, the reasons for discarding differ. Specifically, in developing countries, cereals are likely to be discarded due to a lack of facilities for processing postharvest crops, which is considered food loss. Conversely, in developed countries, cereals are discarded due to consumers' high-quality standards for food and retailers' business strategy (i.e. overstocking of food), which is also considered to be food waste.

3. Process of handling postharvest crops

As discussed in the previous section, the discarding of food in developing and developed countries is mainly due to food loss and food waste, respectively. This section focuses on the case of food waste. Postharvest crop processing involves several steps. First, the postharvest crops are cleaned.[3] This process is usually accomplished mechanically in developed countries, whereas the same processes are done manually in developing countries. Consequently, the manual methods are less efficient than the mechanical means and are only able to process smaller amounts of postharvest crops. Manual work of threshing sometime causes postharvest crops to be in incomplete or damaged conditions, rendering some processed crops to be deemed inappropriate for sale. Second, many types of postharvest crops must be dried, as they originally contain water and cannot be stored for a long time without moulding unless the water content is reduced to an appropriate level. Similar to the process of cleaning, in developing countries, the drying process is usually done manually, whereas postharvest crop drying is mechanically processed in developed countries. In developing countries, postharvest crops are usually dried in the sun, whereas mechanical dryers are used in developed countries. Sun drying is easily affected by weather conditions, takes time and the amount of crops that can be dried at once is limited. Using mechanical dryers is a more efficient means to stably dry a large number of crops. Third, postharvest crops are stored in facilities. Securing postharvest crops in a storage facility is important for the stable delivery of large quantities of crops to retailers. Nevertheless, many grain storage facilities in developing countries are simple structures made from building materials such as wood, grass or straw. During storage, many stored crops are damaged by insects and mould, as no accommodations such as pest, temperature and humidity control are available. Finally, postharvest crops are delivered to retail stores. In developed countries, transportation operations are exceptionally efficient; crops are collected at distribution bases, finely sorted and delivered to various locations by delivery vehicles equipped with temperature and humidity control functions. Such crops are delivered without degradation of quality. Furthermore, most roads in developed countries are paved, making it easy for delivery vehicles to move smoothly and expediently reach destinations. This prompt delivery system ensures that crops are supplied to consumers in good condition. In contrast, the circumstances in developing countries differ widely from developed countries. The means of transportation is not limited to vehicles, but also carried via motorcycles or livestock. Of course, these means of transportation do not have temperature or humidity control functions. In addition, many roads are unpaved, slowing delivery, and vibrations often damage crops.

In addition to the above processes, studies also highlight the lack of market supply chains in developing countries. If producers do not have a dependable, expedient and equitable means of transporting crops to consumers, extensive losses

[3] Some types of postharvest crops such as grains are threshed prior to being cleaned.

can occur [6]. This circumstance is amplified by the lack of communication between producers and consumers [6].

As the above has established, significant differences exist between developed and developing countries in the process of handling postharvest crops. The ironic circumstances of food waste in developing countries suffering from food insecurity reveal that significant inefficiencies in postharvest processing in developing countries are based on a lack of capital and infrastructure.

4. Improvement of postharvest technology in developing countries through recapitalisation

As noted in the previous section, insufficient capital and technology in developing countries cause the wastage of many crops at postharvest stages. This section considers various measures and attending challenges to improvement in developing countries. Specifically, capital investments are essential to the improvement of postharvest crop handling; however, such investments generally require a large sum of money, and producers in many developing countries cannot afford to cover such costs. Therefore, one of the most feasible approaches relies on development support through foundation bodies (i.e. international organisations or wealthy countries). The logical concern of foundation bodies is whether the expected result of support (i.e. reduction of postharvest loss in developing countries) will be sufficient to cover investment costs. If such development support does not adequately reduce the postharvest loss, the foundation bodies will consider the development support to have been unsuccessful. Some studies have noted that, although capital for handling postharvest crops is available, sufficient reduction of postharvest loss is not achieved. According to a study that conducted interviews with participants in Egypt, Indonesia, Kenya, Ghana and India, the simpler the postharvest technology, the better it's chance of adoption, sustainability and long-term use, and the opposite is also true [7]. This is mainly because local farmers cannot fully utilise complicated postharvest technology that requires [7]. Moreover, extension services for farmers are shown to be effective in reducing postharvest losses of rice crops in Bangladesh [8].

It has also been reported that even if the capital is well equipped, the results from support will vary depending on how the farmers grow their crops. In the Ludhiana and Ferozepur district of Punjab, wheat harvesting losses are high during late harvesting due to the shattering of grains; hence, it is asserted that the farmers should be advised to undertake timely wheat crop harvesting to minimise harvesting losses [9]. Moreover, a significant number of horticultural crops in developing countries such as Ethiopia are wasted because most of these crops are produced by small-scale farmers with limited knowledge and financial sources [10].

In developing countries, crop prices are also volatile, and this volatility can affect investment outcomes. When making a capital investment for storage, if a crop is traded at a high price, the return on the investment exceeds its cost; however, if a crop is traded at a low price, the opposite is true and the capital investment is considered unsuccessful [11].

Some studies have argued that proper grading systems for postharvest food will help to reduce postharvest losses. For example, one study argues that better handling, packing and grading are needed to reduce postharvest grape losses in Pakistan, noting that if there are only grade 'A' grapes in a crate, the retailer will obtain much higher net revenue for the same price as multiple crates of poor grade grapes [12]. A research survey revealed that a crate had 17% damaged grapes, on average; out of which 11% was wastage [12]. Some studies have focused on crop packing to reduce postharvest loss. For example, one study endeavoured to design modified atmosphere packaging for postharvest mushroom storage [13]. The proposed packaging alters the normal composition of air to provide an appropriate atmosphere to decrease products' respiration rate [13].

Some studies have attempted to reduce postharvest loss through supply chain development. It was confirmed that rice production capacity in Nigeria significantly improved to ensure food security; however, postharvest rice crops are not sufficiently distributed to consumers due to a lack of a postharvest management system [14]. The authors argue that technologies are available but are not in the hands of farmers and other actors in the rice value chain; thus, the dissemination of existing technologies for managing postharvest crops is required [14]. Another study argues that management of temperature and humidity (i.e. refrigerated transportation, cold storage at wholesale distribution centres, refrigerated retail display and cold storage at home) are essential for reducing postharvest crop losses [15]. The authors conclude that there are three steps to reduce postharvest losses: first, application of current knowledge to improve the handling systems of horticultural perishables (particularly packaging and cold chain maintenance); second, overcoming socio-economic constraints, such as infrastructural inadequacies and poor marketing systems; third, encouraging consolidation and vertical integration among producers and marketers of horticultural crops [15].

As noted, postharvest losses in developing countries occur due to a lack of capital investment for infrastructure to handle postharvest crops; however, simply providing capital does not always reduce postharvest losses, as there are multiple issues to consider. For example, if the installation accommodated by capital is not fully utilised, reduction of postharvest loss will not be achieved. To do so, local producers must be trained to make full use of the equipment and supply chain improvement is also necessary.

5. Theoretical analysis of capital investment for handling postharvest crops

This section presents the economic theoretical model illustrating the circumstances of capital investment for managing postharvest crops in developing countries. Because many of this chapter's readers are assumed to be unfamiliar with these constructs, a simple theoretical model was formulated. As established, the development of facilities with sufficient functions is necessary to reduce postharvest losses; however, budget constraints render many developing countries unable to afford the construction of such facilities. Subsequently, development support from foundation bodies (i.e. international organisations and investors) presents a realistic financing method. To encourage this development support, sufficient outcomes (i.e. reduction of postharvest losses) are required; however, as discussed in Section 4, some capital investments for managing postharvest crops in developing countries are not achieving the expected results (i.e. postharvest loss has not decreased as much as expected). Additionally, we also provide case studies of agricultural development support from a government organisation, a non-government organisation (NGO) and foreign direct investment to the developing countries. The case studies are compared to the theoretical analysis to consider the factors behind its successes.

5.1 Illustration of circumstances in which invested capital is not fully utilised to handle postharvest crops

This subsection introduces a theoretical illustration of the circumstances in which invested capital is not fully utilised to reduce postharvest losses. This circumstance can be illustrated by the following setup.[4]

$$Y = L^{\alpha} K^{1-\alpha} \tag{1}$$

where Y is postharvest crops delivered to consumers without being discarded, L represents the number of labourers handling postharvest crops, K is the capital invested to handle postharvest crops and α measures the dependence on L to produce Y.[5] Eq. (1) indicates that postharvest crops are handled by L and K and a higher Y indicates less postharvest losses.

By differentiating Y with respect to L results in the following equation:

$$\frac{\partial Y}{\partial L} = \alpha \left(\frac{K}{L} \right)^{1-\alpha} \tag{2}$$

Eq. (2) measures how much Y increases when 1 unit of L is added. This equation has some notable features. According to Eq. (2), $\frac{\partial Y}{\partial L}$ is positively related with $\frac{K}{L}$; thus, the larger the size of K and the smaller the size of L leads the larger the size of $\frac{\partial Y}{\partial L}$. This implies that when there is an insufficient number of labourers and substantial capital, some capital will be unused. In such circumstances, additional labour will significantly improve productivity. In contrast, if there is insufficient capital and a large number of labourers, additional labourers will be unable to leverage capital to produce Y. Furthermore, if the postharvest crop handling depends excessively on labourers, α will be close to 1; hence, $\frac{\partial Y}{\partial L}$ becomes a larger value. Conversely, if the postharvest crop handling depends excessively on capital, α becomes close to 0, then $\frac{\partial Y}{\partial L}$ becomes close to 0.

By differentiating Y with respect to K, the following equation is generated.

$$\frac{\partial Y}{\partial K} = (1-\alpha) \left(\frac{L}{K} \right)^{\alpha} \tag{3}$$

According to Eq. (3), $\frac{\partial Y}{\partial K}$ is positively related with $\frac{L}{K}$; thus, a larger size of L and smaller size of K leads to a larger size of $\frac{\partial Y}{\partial K}$. Moreover, if α is close to 0, $\frac{\partial Y}{\partial K}$ becomes the larger value. Conversely, if α is close to 1, $\frac{\partial Y}{\partial K}$ will become close to 0.

Considering the relationship between α, $\frac{\partial Y}{\partial L}$ and $\frac{\partial Y}{\partial K}$, the production process appears to depend excessively on one input (i.e. L or K), and adding a few dependent inputs (i.e. L in the case of small α and K in the case of large α) does not increase production as expected.

Due to lack of sufficient knowledge and skills, even if high-performance equipment for managing post-harvest crops is installed, farmers are unable to fully utilise it. This circumstance reflects a case of α almost equalling 1. As was demonstrated, when $\alpha \approx 1$, $\frac{\partial Y}{\partial K}$ becomes close to 0. Thus, capital investment for the management of postharvest crops does not significantly improve the situation. To improve this circumstance and sufficiently reduce the loss of postharvest crops with the capital

---

[4] Eq. (1) takes form of Cobb–Douglas production function, which is often used by economists to express output as a function of labour and capital.

[5] We assume that the number of crops harvested is fixed.

investment, production must be processed with a smaller value of α. This is only possible when labour is capable of utilising the capital.

5.2 Funding Body possible when under outcome uncertainties from development support for reducing postharvest losses

As illustrated in Subsection 5.1, if labourers are unable to utilise the capital for handling postharvest crops, increased capital does not reduce postharvest losses. This subsection analyses the process of the funding body vest crops with the capital investments. An additional theoretical setup is introduced as a part of this analysis, partially referring to the equations in the previous subsection. The setup assumes two time periods ($t = 1$ *and* 2).[6] The funding body expects the provision of development support to increase K, which will cost C. If the support is provided at $t = 1$, K is enhanced at $t = 2$; however, improvement of postharvest management depends on how much the labour can utilise the invested capital, which is assumed by two potential values of Y at $t = 2$, represented by Y_1 and Y_2, which satisfy a condition of $Y_2 < C < Y_1$. Y_1 indicates that development support can sufficiently contribute to reducing postharvest losses, whereas Y_2 indicates that the cost of development support exceeds its contribution. A chance of having Y_1 is probability P where $0 \leq P \leq 1$, and P is inversely related to α. Finally, the model uses a risk-free rate r to discount the values at $t = 2$ to $t = 1$.[7] This is illustrated below in **Figure 1**.

A present value of expected outcome $E(O)$ from the development support can be calculated as follows.

$$E(O) = \frac{P \times Y_1 + (1 - P) \times Y_2}{1 + r} \tag{4}$$

As it is clear from Eq. (4), the size of $E(O)$ depends on, and P depends on α. From the funding body's perspective, development support is not worth providing unless it can contribute to sufficiently reducing postharvest losses. To investigate the decision to provide development support, this study uses the return on investment (ROI) shown in Eq. (5).

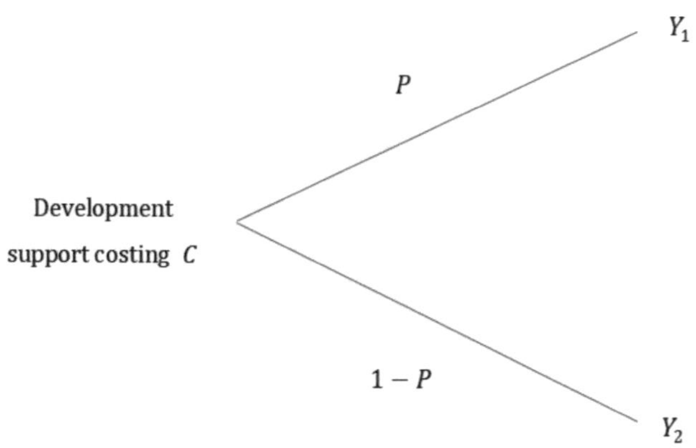

Figure 1.
Outcome from development support for reducing postharvest losses.

---

[6] In this setup, Eqs (1)–(3) are assumed at $t = 2$.

[7] Financial analysis generally considers that a future monetary value is lower than the corresponding current monetary value (i.e. 1 dollar at $t = 2$ is valued lower than 1 dollar at $t = 1$). This is because money at a current time can be invested, and an investment return can be realised at a future time.

$$\text{ROI} = \frac{E(O) - C}{C} \tag{5}$$

The funding body decides to provide development support when $ROI > 0$, and the opposite is true in the case of $ROI < 0$.[8] As is clear from Eq. (5), ROI is positively related to the size of $E(O)$; hence, when α is sufficiently reduced, the development support is provided.

Summing up the above analysis, if labour does not have enough capacity to utilise the invested postharvest technology, the funding body forgoes development support.

5.3 The case studies of capital investment for handling postharvest crops

In Subsections 5.1 and 5.2, we theoretically demonstrated the situation where invested capital is not fully utilised to handle postharvest crops and the funding body's judgement under outcome uncertainties from development support for reducing postharvest losses. In this subsection, we analyse the three successful cases of agricultural development support from the government organisation, NGO and foreign direct investment to the developing countries. We compare the cases and the theoretical model to consider the factors behind the success of the cases.

5.3.1 Post-harvest handling and storage (PHHS) project in Rwanda

First, we try to consider the successful case of the government organisation (U.S. Agency for International Development (USAID)).[9] USAID is managing PHHS Project carried out between September 2009 and August 2013, which budgeted 8.3 million US dollars [17]. The PHHS project is to integrate local farmers into commercial marketing channels as a way of driving investment in postharvest technology and process improvements for maize, beans and rice in Rwanda [17]. According to the report, the agricultural products are fragmented in Rwanda due to a lack of the farmers [17]. According to the report, the agricultural products are fragmented in Rwanda due to lack of the farmers' capital and know-how to efficiency harvest, store and market their surplus yields [17]. This situation is reminiscent of the case of having a high level of α and low level of K in Eq. (1) (i.e. there is a lack of capital, however, because the production process is overly dependent on labours, increasing capital cannot be expected to reduce the postharvest loss significantly). As was demonstrated in Subsections 5.1 and 5.2, one of the measurements to overcome the situation is reducing the value of α in Eq. (1) by improving the skill of local farmers. The latter can be achieved by providing training for local farmers. As a part of the PHHS project, the training for technical and business practices in postharvest handling are provided [17]. The training can be considered to help the local farmers to effectively utilise the capital to manage the postharvest products.

Additionally, we also pay attention to the funding process of this program. PHHS set an objective to mobilise private investment and bank finance to develop businesses that require storage infrastructure [17]. Such funding bodies are like the one illustrated in Subsection 5.2, which dislikes uncertainty of outcomes. If there is no intervention of PHHS, the private funding bodies would be discouraged to fund

---

[8] The funding body can be considered indifferent regarding the choice to provide or not provide the development support when $ROI = 0$.

[9] USAID is established in 1961 to lead the US government's international development and humanitarian efforts [16].

the project, because they can hardly expect its succession. However, the intervention of PHHS helps the private funding bodies to understand the project and expect α in Eq. (1) to be reduced. This is to alleviate the asymmetry of information, which is a barrier to undertaking financial transactions.

As a result of this project, 104 storage centres were constructed or rehabilitated and over 60,000 farmers were trained in postharvest handling and storage best practices [17].

5.3.2 Sasakawa Africa Association (SAA)'s Activity in Africa

In this subsection, we consider the case of the NGO called SAA that provides agricultural development support in Africa.[10] Although SAA is an NGO, they have a large financial resource and can provide sufficient assistance for the agricultural development in African countries [18]. SAA organises its operations under five key themes, and one of them (i.e. Theme 2) is to promote postharvest handling and agro-processing in African developing countries [18]. Their activity is diversified to many countries. For example, SAA launched a project in Tanzania to promote improved grain storage, both on the farm and also in the form of communal storehouses [18]. Moreover, at several locations in Tanzania, organised farmer groups were linked to credit institutions that would provide short-term loans against grain held by the associations in bonded warehouses [18].

The projects of SAA and USAID is in common with emphasising human resource development (training) in addition to providing financial support and capital constrictions between 1989 and 1995, about 1,000 of the country's 5,000 government extension workers received crop management training from Sasakawa-Global 2000 staff. Such training leads to lowering the value of α in Eq. (1).

In addition, rather than providing all the financial support, SAA and USAID tries to act as a bridge for transactions with other financial institutions and makes efforts to attract further investment.

Providing funding support to farmers in developing countries is accompanied by the risk due to the lack of available information and the uncertainty of future outcomes. If a prominent NGO like SAA act as a bridge, investors' anxieties will be alleviated and the problems shown in Subsection 5.2 will be resolved.

5.3.3 China's overseas investment in agriculture

Unlike the case in Sub-subsections 5.3.1 and 5.3.2, some countries encourage private investors to provide agricultural development support to the developing countries for a purpose of meeting their own needs. For example, China has promoted to increase overseas foreign direct investment (OFDI), and a large portion of OFDI is devoted to agricultural support in developing countries [19].

China is one of the major producers of agricultural products in the world; however, the contribution of agriculture to their GDP has dropped considerably over the past three decades [19]. This is mainly due to China's remarkable economic growth that has led to industrialisation. As a result, China faces a challenge in sustaining the self-sufficiency of agricultural products and increased import of agricultural products from other countries [19]. Especially, in recent years, China has emphasised ties with ASEAN, and OFDI to Cambodia, Myanmar and Laos is increasing [19].

---

[10] SAA is the NGO established in 1986 aiming to support the development of farming productivity and the rural economic value chain [18].

In Cambodia, many Chinese firms are directly involved in the development of plantations [19]. As a result, the export of rice, cassava, sugar and cocoa from Cambodia to China has significantly increased from 2012 to 2015 [19]. Similar to the case of Cambodia, the export of maize, rice, beans and oilseeds and tea from Laos to China has significantly increased [19].[11]

Similar to Sub-subsections 5.3.1 and 5.3.2, in this case, a large amount of capital was injected into the agricultural industry in the developing countries through OFDI from China. However, the report also points out some challenges in local work sites. For example, communication and consultation with local people are often challenging for foreign investors, and having lack of information flows causes distrust between foreign investors and local people.

As can be seen from the case studies in this section, there are some successful trials for providing agricultural development support to farmers in developing countries. The trials are operated by many types of organisations (i.e. government organisations, NGOs and private investors). One of the reasons for the success of these trials is that they provide comprehensive support such as training local farmers through agricultural development support. Moreover, by eliminating information asymmetry between investors and local farmers causing concerns about uncertainties of investment, the organisations help local farmers to obtain loans from private financial institutions.

6. A comparison of agricultural investment ratio (AIR) and development flows to agriculture (DFA) between least developed and G20 countries

Section 5 theoretically demonstrated that even if the capital for handling postharvest crops is increased, reduction of postharvest crops will not be achieved unless local labourers can fully utilise it. In addition, we considered the cases that succeeded the agricultural development support in the developing countries by providing comprehensive support such as training local farmers. However, such successful support is limited to the small number of cases, and many developing countries do not receive effective support. This section presents a comparative investigation of data regarding AIR and DFA to overview agricultural developing support in developing countries around the world.

6.1 A comparison of AIR between least developed and G20 countries

This subsection comparatively analyses the AIR values in the least developed and G20 countries. The AIR value is calculated by dividing the 'Agriculture Gross Fixed Capital Formation' (Domain code CS, Element code 6110 and Item code 22030 in FAOSTAT) in country i at time t by 'Agricultural Value Added' (Domain code MK, Element code 6110 and Item code 22016 in FAOSTAT) in country i at time t.[12] The AIR value indicates how much of the total factor income is reinvested in new fixed assets in the agricultural industry of the respective country [20]. **Table 3** demonstrates the AIR value in the least developed countries in 2018 and 2019.[13]

---

[11] Because the situation in Myanmar has changed drastically in recent years, we decided not to refer the agricultural development in Myanmar.

[12] The area and year codes of the variables are not shown because these depend on the countries and years selected. For information on these associated codes, please refer the data source.

[13] Since realisation of outcome from a capital investment takes time, **Table 3** presents the AIR for both 2018 and 2019.

Countries	AIR (2018)	AIR (2019)
Afghanistan	0.06	0.05
Angola	0.1	0.14
Bangladesh	0.09	0.1
Benin	0.04	0.04
Bhutan	0.11	0.13
Burkina Faso	0.07	0.08
Burundi	0.04	0.04
Cambodia	0.09	0.09
Central African Republic	0.04	0.02
Chad	0.06	0.07
Comoros	0.07	0.07
The Democratic Republic of the Congo	0.05	0.05
Djibouti	0.07	0.07
Eritrea	0.22	0.23
Ethiopia	0.07	0.07
Gambia	0.04	0.03
Guinea	0.12	0.13
Guinea-Bissau	0.05	0.04
Haiti	0.06	0.06
Kiribati	0.04	0.04
Lao People's Democratic Republic	0.09	0.1
Lesotho	0.08	0.09
Liberia	0.04	0.05
Madagascar	0.05	0.05
Malawi	0.04	0.04
Mali	0.07	0.07
Mauritania	0.11	0.12
Mozambique	0.05	0.05
Myanmar	0.11	0.1
Nepal	0.06	0.06
Niger	0.02	0.02
Rwanda	0.07	0.07
São Tomé and Príncipe	0.05	0.04
Senegal	0.09	0.08
Sierra Leone	0.05	0.07
Solomon Islands	0.09	0.1
Somalia	0.05	0.05
Timor-Leste	0.11	0.11
Togo	0.04	0.04
Tuvalu	0.08	0.06

Countries	AIR (2018)	AIR (2019)
Uganda	0.08	0.08
United Republic of Tanzania	0.07	0.08
Vanuatu	0.13	0.13
Yemen	0.1	0.1
Zambia	0.11	0.08

Source: FAOSTAT (http://www.fao.org/faostat/en/#data).

Table 3.
AIR in the least developed Countries in 2018 and 2019.

Table 3 demonstrates that most AIR values in the least developed countries are less than 10%. This low percentage of AIR indicates that a small portion of the total factor income is reinvested in new fixed assets in the agricultural industry of these countries. In addition, since no remarkable increase in the AIR is seen from 2018 to 2019, these countries are likely to continuously record low AIR. This indicates that the agricultural capital in these countries is inadequately structured. There are various possible causes for this situation; for example, in developing countries, various areas other than agriculture (i.e. medical care and education) are underdeveloped and are often prioritised, minimising the amount of funding available for agriculture. **Table 4** presents the AIR in G20 countries in 2018 and 2019.[14]

Countries	AIR (2018)	AIR (2019)
Argentina	0.14	0.15
Australia	0.26	0.36
Brazil	0.15	0.15
Canada	0.21	0.2
China, mainland	0.13	0.14
France	0.29	0.29
Germany	0.44	0.36
India	0.13	0.12
Indonesia	0.16	0.15
Italy	0.23	0.2
Japan	0.24	0.25
Mexico	0.03	0.03
Republic of Korea	0.16	0.19
Russian Federation	0.22	0.19
Saudi Arabia	0.1	0.11
South Africa	0.21	0.25

[14] The EU is omitted in **Table 4** for the same reason as in **Table 3**. Moreover, due to the data availability, data for China (mainland) is presented, rather than the whole of China.

Countries	AIR (2018)	AIR (2019)
Turkey	0.16	0.17
United Kingdom of Great Britain and Northern Ireland	0.4	0.42
United States of America	0.38	0.39

Source: FAOSTAT (http://www.fao.org/faostat/en/#data).

Table 4.
AIR in G20 countries in 2018 and 2019.

The data reported in **Table 4** differs from that of **Table 3**. Specifically, all the AIR values in **Table 4** exceed 10% and some exceed 40%. The G20 countries have strong national power and can invest large sums in agriculture. Moreover, the existing facilities are far more substantial than the least developed countries. Furthermore, many of the AIR values in 2019 do not differ largely from those of 2018, indicating that these countries are constantly recording the high value of AIR.

6.2 A comparison of DFA between the least developed and the G20 countries

The least developed countries do not have adequate funding for sufficient capital investment in the agricultural sector. Agricultural capital investment in the least developed countries relies on development support from funding bodies as one of the most realistic means. **Table 5** presents the DFA (in millions of US dollars) from the international organisations to the least developed countries.[15]

Organisation (donors)	DFA (2018)	Number of recipient countries (2018)	DFA (2019)	Number of recipient countries (2019)
Adaptation Fund			19.9	3
African Development Bank (AfDB)	322.6	11	426.8	18
Arab Bank for Economic Development in Africa (BADEA)			19.3	2
Asian Development Bank (AsDB)	331.1	7	669.2	8
Climate Investment Funds (CIF)	14.6	1		
European Union Institutions	409.5	11	127.6	8
Food and Agriculture Organization of the United Nations (FAO)			15.9	46
Global Environment Facility (GEF)	42.1	8	45.6	17
Green Climate Fund	64.8	2	96.8	6
Inter-American Development Bank (IDB)			0.8	1

[15] As this chapter does not afford enough space, detailed information on recipient countries is omitted in **Table 5**. The DFA variable can be identified in FAOSTAT with Domain code EA, Element code 6110, Item code 22041 and Purpose code 310. DFA value is rounded to the first decimal place. The Donor code, Recipient code and Year code of the variables are not presented because these depend on the donors, recipients and years selected. For the information on these codes, please refer to the data source.

Organisation (donors)	DFA (2018)	Number of recipient countries (2018)	DFA (2019)	Number of recipient countries (2019)
International Bank for Reconstruction and Development (IBRD)	1991	48		
International Development Association (IDA)			856.6	20
International Finance Corporation (IFC)			62.4	6
International Fund for Agricultural Development (IFAD)			859.8	36
Islamic Development Bank (IsDB)	364.4	14	201	8
OPEC Fund for International Development (OFID)	105.7	6	164.8	8
United Nations Development Programme (UNDP)	0.6	4	0.6	3

Source: FAOSTAT (http://www.fao.org/faostat/en/#data).

Table 5.
DFA to the least developed countries from international organisations.

There are many international organisations in the world that support economic progress in developing countries. **Table 5** reveals that DFA is allocated by international organisations to the least developed countries; however, there are differences in the number of recipient countries, the size of DFA varies among international organisations and the number of recipient countries and DFA size fluctuates from year to year. Considering the lack of capital of postharvest management, DFA in **Table 5** is unlikely to be sufficient to improve the circumstances. Funding from international organisations is also limited; hence, the provision of substantial DFA to many developing countries is difficult. To improve this support limitation, developing countries should apply strategies for the effective use of capital provided by agricultural development support.

7. Conclusions

Global population growth and environmental burdens have caused rising concerns regarding future food security worldwide. Despite these growing concerns, many crops are discarded without being consumed at postharvest stages. In particular, the postharvest loss is experienced in many developing countries mainly due to a lack of capital and technology for food storage, processing (i.e. threshing, drying and packaging) and distribution. Investigating the existing literature and FAOSTAT data, this study demonstrated that a significant amount of postharvest loss is realised in the least developed countries. Moreover, a theoretical analysis demonstrated that a lack of knowledge and skills among local labourers can result in the lack of full utilisation of improved capabilities for handling postharvest crops, and the postharvest loss is not reduced as expected. Theoretical analyses demonstrated that funding bodies are discouraged from providing support to developing countries in circumstances wherein the outcome of the support due to failing use of capital is uncertain. The proportion of AIR in the least developed countries is significantly lower than that in the G20 countries, which is consistent with the theory of this study indicating the lack of capital for postharvest management in developing countries. DFA to the least developed countries from the international organisations also appeared to be insufficient for improving the current circumstances. Funding bodies also have limited funds, and it can be extremely difficult to provide adequate support. Efforts to maximise the limited capital in developing countries are critical to meaningful improvement.

Acknowledgements

We are deeply grateful for the editor of Postharvest Technology - Recent Advances, New Perspectives and Applications. We are also grateful to IntechOpen and Meikai University, which supported our research in various ways.

Author details

Ryusuke Oishi
Meikai University, Urayasu City, Chiba Prefecture, Japan

*Address all correspondence to: r-o@meikai.ac.jp

References

[1] The United Nations, Department of Economic and Social Affairs Sustainable Development. Topics: Rural Development [Internet]. Available from: https://sdgs.un.org/topics/rural-development [Cited: 6 September 2021]

[2] Commodity Policy and Projections Service Commodities and Trade Division, Food and Agriculture Organization of the United Nations. Trade Reforms and Food Security: Conceptualizing the Linkages. [Internet]. Available from: http://www.fao.org/3/y4671e/y4671e00.htm# Contents [Cited: 19 September 2021]

[3] Kumar D, Kalita P. Reducing postharvest losses during storage of grain crops to strengthen food security in developing countries. Food. 2017; **6**(1):8. DOI: 10.3390/foods6010008

[4] Food and Agriculture Organization of the United Nations. Food Loss and Food Waste [Internet]. Available from: http://www.fao.org/food-loss-and-food-waste/flw-data [Cited: 5 September 2021]

[5] Oishi R. Food loss and waste in Japan. New Food Industry. 2019;**61**:908-914

[6] Kader AA. Increasing food availability by reducing postharvest losses of fresh produce. Acta Horticulturae. 2005;**682**(682): 2169-2176. DOI: 10.17660/ActaHortic.2005.682.296

[7] Kitinoja L. Innovative small-scale postharvest technologies for reducing losses in horticultural crops. Food. 2017; **682**(682):2169-2176

[8] Majumder S, Bala BK, Arshad FM, Haque MA, Hossain MA. Food security through increasing technical efficiency and reducing postharvest losses of rice production systems in Bangladesh. Food Security. 2016;**8**(2):361-374. DOI: 10.1007/s12571-016-0558-x

[9] Grover KD, Singh MJ. Post-harvest losses in wheat crop in Punjab: Past and present. Agricultural Economics Research Review. 2013;**26**(2):293-297

[10] Hailu G, Derbew B. Extent, causes and reduction strategies of postharvest losses of fresh fruits and vegetables—A review. Journal of Biology Agriculture and Healthcare. 2015;**5**(5):49-64

[11] Gorny RJ, Kitinoja L. Capital $ investment in postharvest technology & recovery of invested capital. Perishables Handling Quarterly Issue. 1999;**97**:3-6

[12] Aujla MK, Shah AN, Ishaq M, Fraooq A. Post-harvest losses and marketing of grapes in Pakistan. Sarhad Journal of Agriculture. 2011;**27**(3): 485-490

[13] Ares G, Lareo C, Lema P. Modified atmosphere packaging for postharvest storage of mushrooms. A review. Fresh Produce. 2007;**1**(1):32-40

[14] Danbaba N, Idakwo PY, Kassum AL, Bristone C, Bakare SO, Aliyu U, et al. Rice postharvest technology in Nigeria: An overview of current status, constraints and potentials for sustainable development. Open Access Library Journal. 2019;**6**: e5509. DOI: 10.4236/oalib.1105509

[15] Kader AA. Postharvest technology of horticultural crops—An overview from farm to fork. Ethiopian Journal of Applied Science and Technology. 2013; (1):1-8. DOI: 10.4236/oalib.1105509

[16] U.S. Agency for International Development. USAID History [Internet]. Available from: https://www.usaid.gov/who-we-are/usaid-history [Cited: 10 October 2021]

[17] Dusen VN, Beyard K. Post-Harvest Handling and Storage (PHHS) Project.

Final Report. Kigali, Rwanda: United States Agency for International Development (USAID); 2013

[18] Dowswell C, Melly P, Lewis I, Gavin J. In: Orr P, Gavin O, Jing, Storage P, editors. Sasakawa Africa Association, 4th floor, the Nippon Foundation building 1-2-2, Akasaka, Minato-Ku, Tokyo; 2015:107-0052.

[19] Grimsditch M. Chinese Agriculture in Southeast Asia: Investment, Aid and Trade in Cambodia. Laos and Myanmar: Heinrich Böll Stiftung Southeast Asia; 2017

[20] Liu X, Vollaro M, VanderDonckt M. Food and Agriculture Organization of the United Nations. Agricultural Investment and Capital Stock. FAOSTAT Analytical Brief Series. Vol. 7. Rome: Food and Agriculture Organization of the United Nations; 2020

Section 2

Postharvest Preservation Technology for Field Crops

Chapter 2

Postharvest Preservation Technology of Cereals and Legumes

Theophilus M. Ikegwu, Clement C. Ezegbe, Chioke A. Okolo and Chigozie E. Ofoedu

Abstract

Cereals and legumes are prone to perishability and have very short shelf-life if not given proper treatment. During different handling and marketing operations, there is a huge postharvest loss of agricultural produce. The qualitative and quantitative losses incurred in cereals and legumes commodities between harvest and consumption are huge. Qualitative losses such as loss inedibility, nutritional quality, calorific value, and consumer acceptability of fresh produce are much more difficult to assess than are quantitative losses. The major cause of postharvest loss (PHL) is the availability of poor infrastructure for postharvest technology (PHT) and processing of commodities. These losses can only be minimized by proper handling, marketing, and processing of the agricultural commodities; as well as the use of modern preservation technologies such as irradiation, radio frequency heating, etc. The sufficient knowledge of pre-and post-harvest preservation technologies and the provision of adequate and sufficient storage facilities for cereals and legumes handling and distribution would help to mitigate the incidence of postharvest deterioration and therefore improve the availability of cereals and legumes in the market and subsequent reduction in malnutrition for increased food security. Postharvest preservation technology of cereals and legumes is very fundamental in reducing postharvest losses and increasing food security.

Keywords: postharvest, physiology, deterioration, losses, postharvest technology

1. Introduction

The cereals are monocotyledons while the legumes are dicots. The cereal belongs to the grass family with more than 300,000 species. Furthermore, more than 190,000 species are angiosperms that are economically viable horticultural plants; and there are approximately 50 different types cultivated throughout the world in which about 51 species are grown. However, cereal's contribution to human nutrition cannot be overemphasized as it had been estimated that nearly 18 species of cereals cultivated provide more than 91% of the food supply to the world population. The cereals cover about 74% of the total tilled land surface. It had been estimated that more than 50% of the protein needs of the world population are provided by cereals [1, 2]. Currently, France ranks first in the Export of cereals such as wheat, rice, maize, and barley in Europe but 5th in the production of wheat in the world [1, 2]. Other cereals include millet, sorghum, rye, oats, etc. The major grains such as wheat, rice, and corn add up to make three-quarters of the world-wide production of grain [1, 2]. Therefore, cereal grains remain the main source of dietary carbohydrates for the supply of vital food energy to the diet [1, 2]. Although cereal grains, such as maize, rice, millet, and wheat are mostly in higher demand for energy provision, other cereals also provide very important food uses while there are more researches to explore the underutilized ones [3]. When cereal crops are grown for the edibility of their fruits, they are referred to as *grains* (botanically called *caryopsis*).

Structurally, the cereal seed is composed mainly of two components; the *endosperm* and the *embryo* (*germ*). The endosperm (more than 90% Of the bulk seed) provides the energy. The pericarp (outer wall) develops from the ovary wall and encloses the endosperm. Beneath the pericarp is the testa (a selectively permeable layer) that borders the embryo which is a product of the inner reproductive gland (ovary wall). The permeability of testa to water is high and aids in seed germination but in the presence of salt, the testa may lose its vigor which would consequently lead to nongermination of seeds planted in soils with dissolved salts. The aleurone layer (with thick-walled cells) is free of starch and is the third important layer of cereal grain. Both testa and pericarp are called the bran. Conversely, legumes are flowering plants (dicotyledons) in the Leguminosae family and were derived from the latin word *legere* (to gather) and *legumen* (seeds harvested in pods) during the mid-17th Century. It includes chickpea, black gram, mung bean, and pigeon pea which have an estimated 16,000–19,000 species in 750 genera. Asia ranks first both in area harvested and in production capacity. India, on the other hand, accounts for 75 and 96% of the total global production of chickpea and pigeon pea, respectively [4]. The expression *food legumes* usually mean the immature pods and seeds as well as mature dry seeds used as food by humans. Based on Food and Agricultural Organization (FAO) practice, the term *legume* is used for all leguminous plants. Legumes such as French bean, lima bean, alfafa, or others that contain a small amount of fat are termed *pulses*, and legumes that contain a higher amount of fat, such as soybean and peanuts, are termed *leguminous oilseeds*. Legumes represent an important source of food in developing countries. Soybean, groundnut, dry bean, pea, broad bean, chickpea, and lentil are the common legumes in most countries. In some countries, depending on the climatic condition and food habits, other legumes are grown. Legumes are next to cereals in terms of their economic and nutritional importance as human food sources [3]. They are cultivated not only for their protein and carbohydrate content but also because of the oil content of oilseed legumes such as soybeans.

Legumes are sources of protein and are relatively costlier economically compared to cereals with great food value; and are reasonable nutrients for the maintenance of the body, e.g., vitamins and minerals. The legume has almost the same energy value per unit weight compared to the cereal grains (4.2 kcal), albeit, they provide more calcium, iron, thiamine, riboflavin, pantothenic acid, among others than cereals. The utilization of legumes is highest in India and Latin America owing to religious restrictions and food attitudes. Legumes also contain some anti-nutritional factors, such as trypsin and chymotrypsin, phytate, lectins, polyphenols, flatulence-provoking and cyanogenic compounds, lathyrogens, estrogens, goitrogens, saponins, anti-vitamins, and allergens. However, heat treatment is known to destroy the anti-nutrients, such as protease inhibitors and lectins, although it also destroys vitamins and amino acids. Legumes are a good source of dietary fiber; the crude fiber, protein, and lipid components have a hypocholesterolemic effect.

Healthy cereal grains and legumes are the demanding enterprises of the recent era for the production of high yield in the next season. The cereal grains and legumes must be properly stored for the maintenance of a high-yielding crop. Losses of high magnitude are encountered during storage that is due to biological and non-biological agents. The incidence of high losses of cereals and legumes after harvest in many countries of the world could account for the food security issues such as malnutrition, diabetes, and hunger which are counterproductive to mitigating efforts towards the improvement of food security. The effect of low yield, poor quality of produce, and the prevalence of chemical toxicants and mycotoxin contamination are significant problems that militate against the genuine and concerted efforts to improve postharvest losses (PHLs), provide appropriate handling and processing technologies for improved postharvest opportunities. In an attempt to maintain high-quality crops during postharvest operations (PHOs), care must be taken during harvesting to minimize damage and ensure appropriate postharvest handling techniques. Reliable methods for the assessment of postharvest losses should be developed while the use of the appropriate techniques to minimize loss and ensure the quality and safety of crops that meet quality standards are desired. In developing countries, Nigeria inclusive, cereals and legumes produced mainly by small-scale farmers are produced and stored on farms [4]. Biological and non-biological agents have been implicated in the postharvest losses of cereals and legumes (**Figure 1**) [5, 6].

There is a direct correlation between plentiful harvest and postharvest spoilage. In countries with huge harvests, postharvest losses are higher than in countries with less bumper harvests which may be a consequence of a lack of care arising from a short supply of laborers to preserve the excess grains. Consequently, farmers may be forced to sell their grains at a less reasonable price during the harvesting season to prevent possible postharvest losses. The glut in the price of cereals and legumes could lead to short supply leading to increased losses arising from insect pest attacks (*Prostephanus truncates).* However, the effect of bumper harvests on losses had not been measured, and overall; the effect would be minimal compared with the losses resulting from an unfavorable climate at harvest. Certainly, farmers are often supplied with sufficient storage capacity in developed countries so that at least

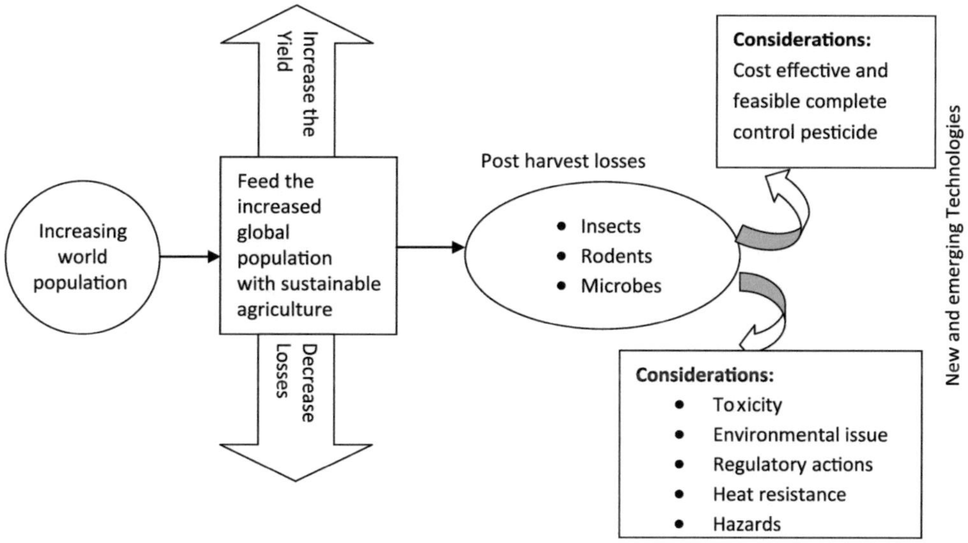

Figure 1.
Considerations for postharvest preservation technologies.

good harvests can be accommodated in fixed stores; unlike in developing countries where less attention is paid to farming and facilities for storage are lacking. In such instances, farmers are content to store surplus cereals and legumes in sacks in their houses. In most cases, especially, in locations where subsistence farming is common, the use of bag storage rather than traditional structures is practiced.

It was strongly believed by the 1970s that postharvest losses (PHLs) at the farm level were high due to traditional practices. However, traditional practices are unlikely culprits as farmers have survived more difficult conditions over long periods by adapting their practices to the situational challenges [7]. Nonetheless, compelling losses do sometimes occur that could be due to agricultural developments for which the farmer is not versed due to nonavailability of extension agents. Among these agricultural developments is the introduction of high yielding varieties that are more susceptible to pest damage, additional cropping seasons that result in the need for harvesting and drying when weather is damp or cloudy or farmers producing significant surplus grains, and because it is to be marketed rather than consumed by the household, the farmer failed to provide the necessary storage facilities for the preservation of the surplus grains.

2. Preservation

Theoretically, any method of food preservation should prevent all the above three (microbial, enzymatic, and proteolytic) types of spoilage. However, current industrial innovation methods have failed to meet these expectations as a whole. Most importantly, microbial spoilage must be prevented at all costs in whichever preservation method was employed, but the effectiveness of thermo-bacteriological treatment for microbial destruction varies to different degrees in the prevention of enzyme activities, proteolytic reactions, and the destruction of different microorganisms. Recent innovative preservation technologies such as ohmic heating, irradiation, infrared, pulse electric field, edible coating, radio frequency, and encapsulation lack the ability to forestall all the concerns posed by spoilage effects completely. These industrial methods employ distinct preservation principles aimed at arresting and or preventing food spoilage. In a nutshell, the industrial method of food preservation makes use of the following principles;

i. The ability to remove moisture through the use of drying/dehydration, evaporation/concentration, etc.

ii. The ability to remove heat from food products by lyophilization/freeze concentration, refrigeration/cold storage, freezing, etc.

iii. Heat addition - heat could be added to food products to destroy microorganisms or inactivate their activities by canning, sterilization, pasteurization, thermization, etc.

iv. Addition of chemicals/preservatives – some chemicals called preservatives may be added to processed food to prevent contamination by the microorganism or forestall enzymatic/browning reactions. Examples of such chemical additives are sorbates, benzoates, etc.

v. Fermentation - during fermentation, secondary metabolites are produced by microorganism which preserves the food product.

vi. Controlled atmosphere storage – in controlled atmosphere storage, the food products' environment is modified to prevent spoilage.

Other methods are the application of high-frequency currents, irradiation, etc. Additionally, other technologies such as pyrolysis, gasification, combustion, and chemical and biochemical processing are used for the conversion of cereals and legumes by-products to chemicals, energy, and other value-added products in the food value chain.

3. Postharvest pest management

Pests pose a very big challenge during the postharvest storage of grain legumes, transportation as well as during distribution. The quality and quantity of grains are reduced by pests if not properly controlled. Pest infestation is a big source of worry for both farmers and food processors because of the losses in investment and profit depletion that come with it. Some of the grain pest control techniques conventionally adopted are fumigation and controlled atmosphere of CO_2 and N_2. Novel techniques have also been developed to take care of some of the shortcomings of conventional pest management practices like fumigation that make use of chemicals. Examples of some of the emerging technologies which have found use in pest management include irradiation, radio frequency, infrared, and microwaves [7]. Methyl bromide application and treatment with hot air on grain legumes storage facilities or systems is also a common practice for disinfection in the grain storage industry [4].

3.1 Irradiation (IR)

Food irradiation is a food preservation technique during which ionizing radiation (0.1–50 KGy) is used to destroy target microorganisms in order to extend the shelf life of foods. During irradiation, microbial inactivation is achieved through free radical development which disrupts DNA and cell membrane integrity [7]. It has shown to be effective in sprout inhibition, elimination of parasites and insects, destruction of spoilage, and pathogenic microorganisms [8].

Radiation treatment at low and moderate doses has been recommended for the disinfestation of legumes [8]. The treatment has also been found to be effective for the reduction of flatulence-causing oligosaccharides as well as trypsin and chymotrypsin inhibitors. With these effects of irradiation on anti-nutritional factors in legumes, the nutritive quality of irradiated beans is thereby improved. Stored produces, especially grains have been successfully decontaminated with ionizing radiation as it affects the internal structure [8, 9]. Irradiation technology has been very effective in controlling the *Aspergillus*, *Penicillium*, *Rhizopus,* and *Fusarium* fungi infection in many grains and prolonging the shelf life over 6 months [10]. The source of radiation that is usually utilized is Co-60 and selenium.

3.2 Radio frequency (RF) heating

Radio frequencies (RF) are electromagnetic waves that are able to penetrate dielectric materials. They usually are characterized by a wavelength of about 11 m and with a frequency range of 1 to 300 MHz [11]. With this ability to penetrate dielectric materials like food grains, they are able to produce heat volumetrically. They are able to do this through ionic polarization or dipole rotation. With the higher moisture content in food grains, their ability to act as dielectric materials is increased, allowing them to act as electric capacitors and resistors and useful in the storage and conversion and electrical to thermal energy. This can be possible within an electromagnetic field [11].

In comparison, the higher moisture content in insects and the consequent higher electrical conductivity would make them require higher lethal temperatures and higher lethal time. At a lethal temperature and time of 50°C for 29 minutes or 54°C for 5 minutes, it would be possible to completely destroy a wide range of insects. This process of higher heating rates and its application finds use in the disinfection of grains on an industrial scale [9, 11].

When insects feed directly on grains, they produce webs and feces on stored pulses thereby reducing grain quality and this represents a huge challenge during the storage, transportation, and distribution of grains. To mitigate this huge challenge, RF heating has been used in the disinfection of dried cereals and legumes. This was demonstrated using a 27 MHz and 6 kW RF unit where the RF proved superior to forced hot air with respect to heating time required (5–7 minutes as against 275 minutes) to heat 3 kg of legumes to 60°C. Good quality product and uniformity in temperature distribution across the surface and interior of the legumes was achieved in the legume samples by a combination of RF heating followed by a movement of forced hot air as grains move through conveyors at 0.56metres per minute. The final interior temperatures of the containments used were above 55.8°C while 57.3°C was recorded for the surfaces of all legumes tested with resultant low index values for uniformity of 0.014–0.016 (ratio of standard deviation to the average temperature rise) for the distribution of interior temperature and 0.061–0.078 for the distribution of surface temperatures. Legumes treatment with RF in combination with forced hot air (60°C) to retain the needed treatment temperature for 10 min followed by the rapid cooling of the air through a 1 cm product layer yielded products with high quality. There were no significant differences in weight, moisture, color, and germination when samples used for control were compared to treated ones [12].

3.3 Infrared

Infrared is a segment in the electromagnetic spectrum found in between the microwave region and the visible spectrum area characterized by a wavelength of about 0.5 to 100 μm [9]. The absorption of infrared rays produces vibrations in the molecules of water, with consequent heat generation. Infrared-based technologies have been found to be energy-efficient and eco-friendly when compared with other conventional methods. Infrared technology also has many other merits like short process duration, uniform effect on food material, low energy requirement, high rate of heat transfer, and enhanced quality of products [9]. As a result of some of the above-listed characteristics, infrared-based technologies have been used in very many food operations like boiling, heating, drying, peeling, recovery of polyphenols and antioxidants, freeze-drying, roasting, microbiological inactivation, grains sterilization, juice and bread production, and cooking. The idea of the usage of infrared rays to disinfect/sanitize grains was established in the early 1960s and 1970s. Based on its exceptionally effective microbial inactivation characteristics, grain industries usually adopt it as a preferred operation for grain disinfection against various chemical methods. Infrared operations involve three different mechanisms in destroying micro-organisms namely thermal inactivation, induction heating, and the distortion of DNA integrity. As documented by [9], the Infrared treatment of mung bean for 5 minutes at an intensity and temperature of 0.29 kWm and 70°C respectively resulted in the total inactivation of fungal growth. Since the penetration rate of infrared is low, its effectiveness gets reduced with an increase in the depth of food. It is therefore recommended more for food surfaces sterilization than other processes. Catalytic-infrared emitters have also been developed and used for the control of weevils in rice, merchant grain beetle, and saw-toothed grain beetle. Generally, a little exposure of about 60 seconds is adequate to destroy insects that strive externally or internally in the grains kept in storage facilities [9].

3.4 Microwaves

Microwaves are electromagnetic radiations with short-wavelength; which has an excellent microbial destruction potential when compared to other conventional chemical methods. Microwave technology is now a highly adopted process by most grain industries for disinfection [8, 13]. They provide protection on grains from insects [10], storage fungi, and field fungi [12]. However, treatment with the use of microwave can induce several adverse effects on seed germination and can affect grain quality. These adverse effects of microwaves are due to variations in heating caused by the difference in cold and hot spot temperature [9].

3.5 Fumigation

Fumigation is a very active pest control technique. Phosphine gas for example is used to kill grain pests at every growth level of their life cycle; this is inclusive of pests with high resistance ability. Nonetheless, the phosphine gas application level needs to be up to 300 parts per million (ppm) and sustained at this level for a minimum of one week at 25°C or more. Alternatively, at a temperature of 25°C or less, a 200 ppm concentration of phosphine gas should be maintained for 10 days for effective and efficient destruction of pests that destroy legumes. Phosphine application exists in two forms; they include bag chains and tablets. There are also a number of ways with which each choice can be adopted effectively in a gas-proof secured silo. Bag chains are also considered a very safe system that assures one of not having any fumigant residue on the grain nor having the operator harmed in whatever way. The next form that phosphine exists in tablet form and is the most widely used and accepted. There exists a third approach in phosphine application which involves the use of a phosphine blanket and is mostly used for very large storages of above 600 tones. The application of phosphine and the concentrations to be used depend on the silo (which should be gas-tight and sealable) volume used for the fumigation. The phosphine concentration to be used is strictly determined based on the volume of the silo rather than the quantity of grain in the silo [13].

An airtight-covered silo especially one that passes the half-life pressure test must have to remain sealed through the entire fumigation period in other to attain a perfect fumigation result with the use of phosphine tablets and/or bag chains. In an airtight-sealed silo, fumigation is expected to last for 7 days with a temperature of above 25°C, and 10 days if the temperature falls between 15 and 25°C. Nonetheless, if the temperature in the silo is less than 15°C, pests particularly insects will be inactive and phosphine is not usually effective at such low temperatures. Based on the ineffectiveness of phosphine at temperatures lower than 15°C, phosphine application is not advisable at temperatures lower than 15°C. The silo must remain closed when fumigation is on and should only be accessed by personnel with suitable personal protective equipment (PPE) as it is dangerous for the operator. Constant opening of the silo is also detrimental to the effectiveness of the fumigation process considering the fact that the phosphine gas concentration and absorption rate would have been reduced below the lethal level recommended for pests' destruction. Recommendations for the phosphine label came to be as a result of detailed testing by the industry, in other words, making use of phosphine as indicated on the label will ensure perfect results [13]. Phosphine is rated high as a very reliable fumigant for the control of pests in grain storage facilities and other production enterprises [13]. Nevertheless, there has been a continuous misuse of fumigants with a resultant effect of poor pest control and the development of resistance in certain species of pests. More so, just as the continuous use of herbicides that has the same principle of action advances weeds being resistant, continuous use of phosphine could lead to grain pest resistance. Nonetheless, in the case of herbicides, the development of resistance by pests can yearly be circumvented by alternating the chemicals used. The same cannot be said for stored grain fumigation as options are limited and where available, they are not cost-effective [13]. In other words, it is best to avoid the resistance of phosphine by using it as instructed.

Other fumigants and a controlled atmosphere may be used for stored grain pests but they are often high in price. However, to prevent resistance of stored grain pests, phosphine sealed in a silo that is impermeable to gas should be used.

3.6 Controlled atmosphere

In spite of the fact that phosphine is the common most used gas fumigant, there exist other gas fumigants for controlling pests in stored grain. These alternatives are however more expensive than phosphine and still require a gas-tight, sealable silo but they offer other options for resistant pest species. Nitrogen (N_2) and carbon dioxide (CO_2) have the advantage of being nonchemical control alternatives. Because nitrogen and CO_2 methods of control change the balance of natural atmospheric gases to produce a toxic atmosphere, they are hence referred to as controlled atmosphere (CA) [13].

3.6.1 Carbon dioxide (CO_2)

Treatment with CO_2 involves displacing the air inside a gas-tight silo with CO_2 at concentrations high enough to be toxic to grain pests. This requires a seal impermeable to gases, measured by a half-life pressure test of no less than five minutes. In order to eliminate all life stages of the main grain pests, CO_2 must be retained at a minimum concentration of 35% for 15 days [14]. To achieve a 35% concentration level of CO_2 for 15 days, 30 kg (size G) cylinder per 15 tones of storage capacity is required. CO_2 is an odorless, colorless, non-flammable gas that is approximately one and a half times heavier than air. Food grade CO_2 comes in form of a liquid in pressurized cylinders and when released from the cylinder, changes to a gas. Carbon dioxide is less effective at temperatures below 20°C. This is because insects are less active at this temperature, so the CO_2 concentration must be maintained for an extended period.

3.6.2 Nitrogen

Grains stored in a nitrogen saturated environment ensure the control of insects and preserve product quality without the use of chemicals [13]. Nitrogen-based storage systems maintain the quality of canola and pulses through the inhibition of the respiration process that causes oxidation, which may result in the increase in free fatty acids, loss of color, and seed deterioration [13]. Grain treatment with nitrogen (for the purpose of pest control) is safe, environmentally friendly, and involves the usage of electricity for its major operations. Nitrogen produces no residues when used, so grains can be sold instantaneously whenever decided as against what is practiced for chemical fumigants which have recommendation period for withholding after fumigation [13]. The use of nitrogen as an insect control technique involves the use of Pressure Swinging Adsorption (PSA) technology in adjusting the atmospheric composition of the grain storage system to expel other gases other than nitrogen, thus depriving the pests of the needed oxygen. The method of application entails purging the silo to its base with gas majorly composed of nitrogen. This is done in order to force out from the silo the oxygen-rich air through the top of the silo. Several hours of operation are required for PSA to build up about 99.5% pure nitrogen and before the air composition reduces to 2% oxygen. It is difficult for adult insects to thrive in 2% concentration of oxygen, provided this concentration is maintained for 21 days at 25°C or above for the temperature of the grain [14]. The inhibition of the different stages of the life cycle of insects (eggs, larvae, and pupae) will be difficult below these recommended temperatures and the number of days for grain storage. For grain temperatures below 25°C, this treatment duration should further be extended to a 28-day period. Additional purging of the silo may be needed to get rid of oxygen that has diffused from the grains and it must be re-evaluated 24 hours after fumigation in order to achieve effective and efficient pest control.

4. Drying technologies

Scientists from all over the world continuously search for new and effective means and use of renewable sources of energy as a result of the continuous increase in the price of fossil fuels and increased levels of greenhouse gas emissions. The world's energy intake is doubled every 20 years and this increase in energy consumption, has resulted in fossil fuels causing many environmental problems and pollution [15]. Drying is a processing technique used for food product preservation and reducing food spoilage. About 3.62% of the world's energy is used for the drying of agricultural products [16].

Presently, the requirement for new drying technology that promotes the higher quality product and efficient drying in shorter periods is the current need. And as a result, hybrid drying systems have emerged as an excellent technique for their versatile drying outcomes, with lower energy requirements and minimum environmental impact. Lately, various hybrid solar dryers which are more efficient in conjunction with other sources for heating the air, hence reducing drying cost and energy consumption have been developed [17, 18].

Grain legumes are usually dried after harvesting before storage in storage facilities [17]. Drying grain legumes to a recommended safe moisture level is fundamental in achieving safe storage of grain legumes. However, too rapid drying of nuts can lead to hardening of the grain core with poor interior while very slow drying may result in microbial growth which will lead to quality deterioration. Recirculation of the solar drying air is thus employed to make efficient use of the heated air by giving a drying rate that provides acceptable product quality.

Drying of pulses is essential because they contain high moisture content of about 18–25% at the time of harvest and, for safe storage, the optimum moisture content need to be in the range of 9–12% to avoid mycotoxin production. It is essential that the grain is dried to a safe moisture level as quickly as possible to avoid deterioration regardless of the drying system employed. There are several techniques of non-natural open-sun drying of grains with hot air. Some of these forms of drying include spouted-bed drying, fixed bed drying, moving bed drying, fluidized-bed drying, and thin-layer drying [19]. Apart from some of these specialized dryers used for grain drying, all-purpose grain drying systems can as well be used in the drying of grain legumes. Generally, as documented by [20], dryers or drying systems are categorized depending on the following:

a. The flow of grain wherein the dryers are denoted as - batch, recirculating and continuous dryers,

b. The relative motion of the grains and the circulating air used for drying. Concurrent, counter-current, cross/mixed flow dryers are found in this category.

c. The source of heat: solar, propane, and electrical dryers are examples of dryers in this category.

Regardless of the type of dryer used in drying grains, the concurrent heat energy transfer and moisture loss principle/process is the same for the drying of grain legumes and equally for other grains [19]. The process of drying grains involves the loss of free moisture which involves the drying of the grain until its equilibrium moisture content is attained. The equilibrium moisture content of the grain implies the final moisture content attained by the grain at a pre-determined relative humidity and temperature. The cardinal factors that influence the drying rate of grain legumes are temperature, grain moisture content, relative humidity, and air velocity [19].

The use of solar dryers is also another medium for drying legumes. A lot of solar drying systems exist for grain drying such as direct, non-direct, and solar. Solar dryers have the problem of the dehydration process being stopped as a result of an absence of solar radiation and absence of radiation at night or low insulation, which decreases the quality of the grains. So far, there have been efforts to proffer solutions to the problems of solar systems, some of which include – the addition of thermal storage materials, phase change materials, and adding a variety of heating modes either direct or indirect [21]. This has led to the evolution of several types of solar dryers. Thermal storage materials have the ability to store thermal energy when there is solar radiation and then make use of this thermal energy when the sun is not available. Three main forms of solar dryers exist with varying sizes, designs, and magnitude [22].

4.1 Classification of solar dryer

The three major types of solar dryers with various sizes, capacities, and designs are:

i. Direct solar dryers

ii. Indirect solar dryers

iii. Mixed-mode solar dryers

4.1.1 Direct type solar dryer

This is a form of the solar dryer where the radiation from the sun is used directly incident on the grains to be drained. The dryers are quite simple in structure, less expensive, little or no maintenance needed, and also simple to use. It can be fabricated with a wooden box with a glass cover and some holes for air entrance and exit also. After the usage of the direct type of solar dryers, the food products are usually not very nice in appearance, color, texture, and with a reduced nutritional quality. In direct-type solar drying, produces to be dried are spread on the ground or mats exposed directly to the sun to absorb solar radiation. As noted by [23], sun-dried grains are prone to high crop losses due to:

i. Non-uniform moisture loss

ii. Attack by insects and rodents

iii. Inability to attain moisture levels that are safe for the safe storage of grains

iv. Proliferation of micro-organisms and possible toxin production.

These challenges have led to the development of other drying techniques like solar drying to overcome the aforementioned challenges. Solar dryers have faster rates, better efficiency, more hygienic with less crop losses when compared to sun drying [23].

4.1.1.1 Open sun drying

Here, food products are placed right under the sun, below solar radiation to get rid of their moisture properties. The difference in density in the air from the atmosphere allows for air movement. In other words, to get a product dried, they are usually spread in a large area under solar radiation. It is usually time-consuming till it is dry to a required level. All that is needed is a large surface ground done with concrete or a suitable soil area with products laid on them between ten to thirty days depending on a favorable weather situation. This form of drying technique consumes a lengthy time of sun subjection which can sometimes bring down the nutrition level of the products, like sapping off their vitamins.

Open solar drying is a good choice for food drying but comes with a lot of problems such as reduced product quality, adverse effects of rain, moist, wind, animal consumption, and dust [24]. The use of industrial drying comes in as another option which is very expensive. It would need a lot of fossil fuel which will result in air pollution. Nonetheless, the spread and adoption of solar energy is likely to take prominence in the coming years and is without negative environmental effective factors [24].

4.1.1.2 Cabinet type solar dryer

The cabinet form of the dryer is advantageous for preserving smaller food products such as vegetables, pepper, and fruits. It has a roof that is transparent with covers that could be either single or two, made using a black-colored plate cover that serves as an absorbing entity for the storage of energy from the sun. Suitable perforated holes allow for the free flow of air and the removal of moisture.

4.1.2 Indirect mode solar dryer

When it comes to moisture removal and heat transfer, indirect sun dryers differ from direct solar dryers. This style of drier is utilized for quick drying. The atmospheric air is heated in a solar air collector in this dryer, and then this hot air moves towards the drying cabin, where products are kept to dry, and the hot air absorbs some moisture from the drying products before exiting through the chimney.

4.1.3 Mixed mode solar dryer

The term "mixed mode solar dryer" refers to a solar dryer that uses both direct and indirect heating methods. The inlet air is heated at the solar air collector before entering the drying chamber in a mixed mode solar dryer. Some of the drying chamber's sides are composed of glass, which adds to the drying chamber's overall warmth. The product is dried using a combination of hot air and direct sunlight in this procedure. In comparison to direct and indirect solar dryers, mixed mode solar dryers require less drying time. Biomass has been used in hybrid sun dryers as an auxiliary heat source to keep drying going all night. Cashews, for example, have been dried in these dryers [23].

4.1.3.1 Greenhouse solar dryer

Tent dryers are similar to greenhouse sun dryers. They have vent sizes that control airflow. Board glazing is used on all sides of this type of drying system. The greenhouse drying system provides a higher degree of control when used in conjunction with the appropriate settings. The main benefit of a greenhouse solar dryer is that it can provide alternate heating with charcoal or briquette burners during inclement weather and can also be used at night.

Greenhouse solar dryers are a type of solar dryer that was developed to address some of the issues that open solar dryers face. The greenhouse solar dryer might be created out of polycarbonate sheets in parabolic shapes, with direct current blowers to help with airflow in the dryer, which has a floor area made out of concrete [24]. Solar radiation intensity was observed between 390 to 820 W/m^2.

Greenhouse drying is one of the world's oldest methods of crop preservation. It entails the phenomena of heat and mass transmission. The product's thermal energy is used in two stages. The temperature of the product rises in the first step due to sensible heat, and the moisture in the product vaporizes in the second step due to the provision of latent heat of vaporization [25]. The greenhouse dryer provides a regulated environment in terms of relative humidity and temperature, which is better for crop drying and hence reduces drying time. The essential processes in the construction of a greenhouse system include vaporization. The greenhouse drier provides a regulated environment in terms of relative humidity and temperature, making crop drying more efficient [25].

a. Natural convection greenhouse dryer

b. Forced convection greenhouse dryer

4.1.3.1.1 Natural convection greenhouse dryer

Incident sun energy is passed through the canopy and used to heat the crops in a natural convection greenhouse dryer. The temperature of the crop rises as a result of solar radiation absorption. The thermosyphic effect is used to operate the natural convection greenhouse drier. Humid air is vented through the dryer's chimney or evacuated through an outlet on the top, while warm air is pumped through the crop by buoyancy forces. Natural convection mode refers to this type of airflow within the drying chamber, and a natural convection greenhouse drier is one that works in this manner [25].

4.1.3.1.2 Forced convection greenhouse dryer

The forced convection greenhouse dryer was born out of a desire for increased air circulation and drying rates. To adjust temperature and moisture evaporation according to the weather conditions, an optimal airflow should be given in the greenhouse drier during the drying process. An exhaust fan on the upper half of the west wall is used to evacuate humid air. Forced convection greenhouse dryers employ a fan or blower to control airflow [25].

The mixed mode solar dryer outperforms other types of solar dryers in terms of drying efficiency, drying time, and thermal efficiency. It has been discovered that a mixed mode solar dryer with a Phase-Change-Material is the best for drying grains with higher efficiency and shorter drying times, as well as being smaller, having fewer moving parts, and requiring less maintenance [19]. In a mixed mode solar dryers with 1.5 m/s air velocity, beans with up to 60% moisture content can be reduced down to 6% within six hours of drying [19]. The time required for drying depends on factors like solar radiation, ambient condition, and relative humidity while the solar collector efficiencies can be as high as 61.82% [21].

5. Storage of grain legumes

Cleaning of grains to remove extraneous materials and contaminants is very fundamental in achieving good and safe storage. It established that cleaning before storage of grains influences the quality of the grain [26]. Cleaning involves the removal of unwanted extraneous material (straws, sand, stone, etc.) from the grain. The storage of grain legumes is a very cardinal stage in the postharvest handling of legumes. Its importance is based on the fact that if the optimal conditions for their safe storage are not maintained a high level of postharvest losses could be incurred. Different microorganisms and pests have the ability to destroy grain legumes after their harvest, during storage, or transportation to various locations of interest. Depending on the prevailing intrinsic and external factors, postharvest losses of grain legumes are estimated to be about 9% for USA and 40–50% for many developing countries [27].

The rapid decline in color, oil quality and ability to germinate, and many other changes in the quality characteristics of grain legumes can be caused by increase in temperature and moisture. High moisture content and elevated temperature of grains can lead to the development of molds in the category of *Aspergillus species, Fusarium species, and Penicillium species,* and the production of some mycotoxins such as aflatoxins, ocharatoxin A, and patulin produced by molds. High moisture content and temperature above optimal levels also aid the infestation of different varieties of insects (granari weevil, grain borer, grain moth, grain beetle, etc.) which feed directly on the grains with a resultant effect of the decline in grain quality and quantity. Infestation of grains by fungi results in reduced nutritional quality, reduction in the quality of proteins that synthesize gluten, and the ability of grains to germinate. Other effects include free fatty acid elevation, lowered starch content, increase in total soluble solids, the decline of non-reducing disaccharides and oligosaccharides. The grains can also be charred due to hot spot development and the formation of mycotoxins may occur as a result of fungal contamination creating very big public health issues [7, 17]. Globally produced grains of about 25% are contaminated by toxins from molds – mycotoxins [28]. The aflatoxins with the greatest intoxicating effect, genotoxic and carcinogenic characteristics of greatest concern are B1, B2, G1, G2, and M1 aflatoxins (**Table 1**) [31].

During storage, grain legume pests are capable of destroying up to 33–50% of global produces [27]. This gives an insight on the seriousness of pest infestation and attack on grains if proper control measures are not put in place. The quality degradation which results in loss of the quantity of leguminous grains globally during storage can get up to 60% in some instances [27]. These losses are primarily

Chemical fumigants (phosphine tablets and methyl bromide)	Compound/Mechanism	Effect	References
Chemical Fumigant	Phosphine Tablets and methyl bromide	Toxic to living organisms and humans	[17]
Sensor-based vacuum hermetic fumigation and storage	Hermetic contaminant, very low oxygen concentration	No harmful effect to humans	[29]
Irradiation	Ionizing irradiation (0.1–50 kGy)	Effective in fungal destruction Grain disinfection	[8]
Radio frequency	1–300 MHz up to 11 M wavelength, penetrate dielectric material and produce heat volumetrically	Destroy insects and disinfect dry grains	[30]
Infrared	0.5–100 μm	Vibrations in molecules of water with heat generation	[30]

Table 1.
Postharvest preservation technologies.

as a result of insect infestation, rodents attack, micro-organisms like mold as well as the breakdown in the normal physiology of grains. It's a well-known fact that pathogenic micro-organisms, insects, rodents, and unwanted contaminants are capable of posing health hazards in grains when consumed. In storing grains from leguminous crops, the usage of suitable packaging and packaging materials is very crucial in achieving good results in postharvest management of leguminous grains. Packaging also serves a very key role during distribution and marketing (to maintain quality) [27].

In village areas of developing and even developed nations, grains including pulses are still kept in traditional storage facilities which are fabricated with natural materials or woven threads. Typical examples of some of the traditional storage structures used include underground pits, thatched roof storage, plastic containers, and basket silos. Though these local structures have a low construction and maintenance cost, they are not very durable, easily invaded by insects and pests resulting in grain legume quality deterioration. Developing nations are currently adopting warehouse storage structures for storing their grains in very large quantities [17].

The materials used for the construction of storage facilities and structures have a direct influence on the moisture content and temperatures in the storage structures [17]. Wooden sticks, concrete blocks, cement, bamboo, and metals (aluminum or steel) are some of the very common materials used for the fabrication of storage structures for grains.

5.1 Silo

Silos are currently very common storage facilities for storing grain in many countries and constitute about 79% of all on-farm grain storage facilities in Australia. Silos are very ideal storage alternative for grain legumes (pulses) especially the cone-based variant which makes for very easy grain unloading/discharge with very low seed damage possibilities [15]. For long-term storage of above three months duration, there is a need for the incorporation of aeration cooling systems and the use of gas-tight sealable storage which are recommended for efficient and effective fumigation regimes in managing and achieving best quality control. Metal silos are fabricated by incorporating augers and ventilators for grain aeration in order to reduce the formation of hot spots. Metal silos with ventilators and augers are considered advanced grain storage systems as they have the ability to extend the shelf-life of grain legumes through controlled respiration and the development of unfavorable conditions for all sorts of grain legume pests [7, 17].

It is advisable to always fill and empty silos from apertures provided at the center of the silos. This is especially important with grains as most grains have a high bulk density and loading or unloading outside the central opening at the center will put an uneven load on the structure which may cause the silo to collapse [14].

Metal silos of different sorts, fabricated with galvanized iron or recycled oil drums have been developed as an economic, effective, and efficient containers-storage option. These silos are suitable for a long duration of storage of cereals and grain legumes in a water-resistant and hermetically controlled environment. Grains stored in metal silos provide protection from rodents, insects, and water, and are thus very good storage systems for pulses [32]. However, there is a need to protect or shield silos from direct sun rays and other heating sources capable of increasing the temperature of the grains contained therein to avoid condensation. As an alternative, silos can be situated in well-ventilated areas with shade to avoid elevating the temperature of the silos [32]. It is worthy of note that metal silos are very efficient and effective for grain storage but they are also expensive [33].

If there is direct exposure of silos to sunlight or the external air is lower than that in the silos which contain the grains, there may be a formation of currents of convectional flow. As a result of the convectional air currents generated, the moist air is being blown pass through the grains. As the moist air travels and meets cooler surfaces like the silo walls, condensation of the moisture will take place and the grain within that area will get dampened. This dampening occurrence is a cardinal problem associated with grains stored in silos made of steel and particularly utilized for storage in hot areas with daily clear sky [28]. High day temperatures and cool night temperatures are a result of a clear sky. The problem of elevated temperatures can be mitigated in small silos by providing a shield in form of a roof or a hat, to prevent direct contact of sun rays with the surface of the silos. Solutions for larger silos may involve grain silo ventilation or transferring of the grains from the silo with a high temperature to another one that has a cool condition. Grain movement during the transfer of grains to another cool silo has the tendency to provide grains with more homogenous moisture content. In a case where the moisture content is too high, then there will be a need to dry the grains again [28].

5.2 Hermetic bags/cocoon

It's still possible for foreign pests like *Callosobruchus maculatus* and *Callosobruchus chinensis* to be located in grain legumes storage systems during storage if appropriate pest management regimes are not strictly adhered to. Grain legumes storage in hermetic bags/Cocoons has to a large extent aided farmers in many countries in storing and extending the shelf life of their produces as they await periods with better produce value and pricing. This has resulted in better financial gains for farmers that make use of Hermetic bags/cocoons storage in extending the shelf life of their produces with the target of a better sales period [33]. The technique of using hermetically sealed polyethylene silo bags is an effective alternative for the protection of stored grain legumes in commercial storage systems and is presently gaining more prominence for both on-farm sites and off-farm sites [34].

6. Conclusion

Legumes are very important food crops that supply good amounts of plant source protein to our meals. Postharvest losses are incurred if grain legumes are not properly handled, prepared, and stored. Some of the notable postharvest handling practices adopted to preserve and extend the shelf life of legumes include drying, pest control, and storage.

Pest control in harvested grains can be achieved through emerging technologies like irradiation, radio frequency ionization, infra-red, and microwave technology. Pest management can also be done through the age-long chemical means of fumigation as well as controlled atmosphere technology as an alternative.

The drying of grain legumes through the traditional means openly spreading in the sun yields poor drying results. Drying of grain legumes is better done through artificial means with hot air dryers or solar dryers of different sorts. Solar dryers have evolved greatly as a result of the need to reduce the level of greenhouse gases emitted by non-solar dryers, high fuel prices to run non-solar dryers, and the need for a renewable type of energy, unlike the non-solar dryers.

Storage of grain legumes for bulk commercial purposes is done in silos while hermetic bag storages are utilized for small-scale storage in other to achieve a fairly optimal storage condition for grain legumes.

Author details

Theophilus M. Ikegwu[1]*, Clement C. Ezegbe[1], Chioke A. Okolo[1] and Chigozie E. Ofoedu[2]

1 Faculty of Agriculture, Department of Food Science and Technology, Nnamdi Azikiwe University, Awka, Anambra State, Nigeria

2 School of Engineering and Engineering Technology, Department of Food Science and Technology, Federal University of Technology Owerri, Imo State, Nigeria

*Address all correspondence to: tm.ikegwu@unizik.edu.ng

References

[1] FAO. World Food Situation. Netherlands: Food and Agriculture Organization of the United Nations; 2021. Available from: https://www.fao.org/worldfoodsituation/csdb/en/

[2] Beverly RL. Safety of food and beverages: Cereals and derive products. Encyclopedia of Food Safety. 2014; **3**:309-314

[3] Stoskopf NC. Cereal grain crops. Reston: Reston Publishing Company, Inc.; 1985

[4] Yousaf Z, Saleh N, Ramazan A, Aftab A. Postharvesting techniques and maintenance of seed quality. In: Araujo S, Balestrazzi A, editors. New Challenges in Seed Biology - Basic and Translational Research Driving Seed Technology. London: IntechOpen (Internet); 2016. Available from: https://www.intechopen.com/books/5218

[5] Delouche JC, Caldwell WP. Seed vigor and vigor tests. Proceeding of Association of Official Seed Analyst. 1960;**50**:124-129

[6] Gregg BR, Billups GL. Seed Conditioning Technology Part A. In: Seed Quality. Vol. 2. USA: Science Publishers; 2010. pp. 1-2

[7] Greeley M. Pin-pointing postharvest losses. Cereals. 1982;**15**(1):30-37

[8] Mohapatra D, Kumar S, Kotwaliwale N, Singh KK. Critical factors responsible for fungi growth in stored food grains and non-chemical approaches for their control. Industrial Crops and Products. 2017;**108**:162-182. DOI: 10.1016/j.indcrop.2017.06.039

[9] Ling B, Cheng T, Wang S. Recent developments in applications of radio frequency heating for improving safety and quality of food grains and their products: A review. Critical Reviews in Food Science and Nutrition. 2020; **60**(15):2622-2642. DOI: 10.1080/10408398.2019.1651690

[10] Hallman GJ. Control of stored product pests by ionizing radiation. Journal of Stored Products Research. 2013;**52**:36-41

[11] Vadivambal R, Jayas DS, White NDG. Wheat disinfestation using microwave energy. Journal of Stored Products Research. 2007;**43**(4):508-514. DOI: 10.1016/j.jspr.2007.01.007

[12] Knox OGG, McHugh MJ, Fountaine JM, Havis ND. Effects of microwaves on fungal pathogens of wheat seed. Crop Protection. 2013;**50**:12-16. DOI: 10.1016/j.cropro.2013.03.009

[13] Ramaswamy R, Krishnamurthy K, Jun S. Microbial decontamination of food by infrared (IR) heating. In: Demirci A, Ngadi A, editors. Decontamination in the Food Industry. 1st ed. USA: Woodhead Publishing; 2012. pp. 450-471. DOI: 10.1533/9780857095756.2.450

[14] Australian Government Grains Research and Development Corporation - GRDC. Vetch Section 13 Storage: How to Store Vetch on-Farm, Aeration during Storage and Stored Grain Pests. Australia: Grain Research and Development Corporation (GRDC). 2018. pp. 1-25. Available from: https://grdc.com.au/__data/assets/pdf_file/0024/370707/GrowNote-Vetch-North-13-Storage.pdf [Accessed: September 13, 2021]

[15] Trostle R. Global Agricultural Supply and Demand: Factor Contributing to the Recent Increase in Food Commodity Price. Outlook Report No. WRS-0801. USA: Economic Research Service, U.S. Department of Agriculture; 2018

[16] Bibin C, Kishore K, Baskar K, Akshay J, Sharma J. Performance analysis of a diesel engine fuelled with Punnai oil methyl ester and its diesel blends. International Journal of Trendy Resources Engineering Technology. 2018;**2**(5):74-79

[17] Sirohi R, Tarafdar A, Gaur KV, Singh S, Sindhu R, Rajasekharan R, et al. Technologies for disinfection of food grains: Advances and way forward. Food Research International. 2021;**145**:110396. DOI: 10.1016/j.foodres.2021.110396

[18] Babar OA, Tarafdar A, Malakar S, Arora VK, Nema PK. Design and performance evaluation of a passive flat plate collector solar dryer for agricultural products. Journal of Food Process Engineering. 2020;**43**(10):e13484. DOI: 10.1111/jfpe.13484

[19] Kundu KM, Das R, Datta AB, Chatterjee PK. On the analysis of drying process. Drying Technology. 2005;**23**(5):1093-1105. DOI: 10.1081/DRT-200059140

[20] Jayas DS, White NDG. Storage and drying of grain in Canada: Low cost approaches. Food Control. 2003;**14**: 255-261. DOI: 10.1016/S0956-7135(03)00014-8

[21] Kumar P, Singh D. Advanced technologies and performance investigations of solar dryers: A review. Renewable Energy Focus. 2020;**35**:148-158. DOI: 10.1016/j.ref.2020.10.003

[22] Sodha MS, Chandra R. Solar drying and their testing procedures: A review. Energy Conversion and Management. 1994;**35**:219-267. DOI: 10.1016/0196-8904(94)90004-3

[23] Tomar V, Tiwari GN, Norton B. Solar dryers for tropical food preservation: Thermophysics of crops, systems and components. Solar Energy. 2017;**154**:2-13. DOI: 10.1016/j.solener.2017.05.066

[24] Nukulwar MR, Tungikar VB. A review on performance evaluation of solar dryer and its material for drying agricultural products. Materials Today: Proceedings. 2020;**46**(1):345-349. DOI: 10.1016/j.matpr.2020.08.354

[25] Chauhan PS, Kumar A, Gupta B. A review on thermal models for greenhouse dryers. Renewable and Sustainable Energy Reviews. 2017;**75**:548-558. DOI: 10.1016/j.rser.2016.11.023

[26] Bucklin R, Thompson S, Montross M, Abdel-Hadi A. Grain storage systems design. In: Handbook of Farm, Dairy and Food Machinery Engineering. New York, USA: Elsevier Incorporated; 2013. pp. 123-175. DOI: 10.1016/B978-0-12-385881-8.00007-0

[27] Uebersax MA, Siddiq M. Postharvest storage quality, packaging and distribution of dry beans. In: Siddiq M, Uebersax MA, editors. Dry Beans and Pulses Production, Processing and Nutrition. 1st ed. New Jersey, USA: John Wiley & Sons, Incorporated; 2013. pp. 75-100

[28] FAO. Grain crop drying, handling and storage. In: Rural Structures in the Tropics: Design and Development. USA: IOWA State University Extension and Outreach; 2018. pp. 363-386

[29] Kumar S, Mohapatra D, Kotwaliwale N, Singh KK. Efficacy of sensor assisted vacuum hermetic storage against chemical fumigated wheat. Journal of Stored Product Research. 2020;**88**:101640

[30] Paul A, Radhakrishnan M, Anandakumar S, Shanmugasubdaram S, Anandharamakrishnan A. Disinfection techniques for major cereals: A status report. 2020;**19**:1125-1155

[31] Eskola M, Kos G, Elliott CT, Hajšlová J, Mayar S, Krska R. Worldwide contamination of food-crops with mycotoxins: Validity of the widely cited "FAO estimate" of 25%. Critical Reviews in Food Science and Nutrition. 2020;**60**(16):1-17. DOI: 10.1080/10408398.2019.1658570

[32] Mobolade AJ, Bunindro N, Sahoo D, Rajashekar Y. Traditional methods of food grains preservation and storage in Nigeria and India. Annals of Agricultural Sciences. 2019;**64**(2):196-205. DOI: 10.1016/j.aoas.2019.12.003

[33] Shendge SN, Pawar VS, Kale PR. Novel technique: Hermetic storage and its application. The Pharma Innovation Journal. 2021;**10**(8):451-456

[34] Silva MGC, Silva GN, Sousa AH, Freitas RS, Silva MSG, Abreu AO. Hermetic storage as an alternative for controlling Callosobruchusmaculatus (Coleoptera: Chrysomelidae) and preserving the quality of cowpeas. Journal of Stored Products Research. 2018;**78**:27-31. DOI: 10.1016/j.jspr.2018.05.010

Chapter 3

Stored Grain Pests and Current Advances for Their Management

Rayees Ahmad, Shafiya Hassan, Showkat Ahmad, Syed Nighat, Yendrambamb K. Devi, Kounser Javeed, Salma Usmani, Mohammad Javed Ansari, Sait Erturk, Mustafa Alkan and Barkat Hussain

Abstract

During the offseason, when fresh food is not available, humans have to consume stored grain food. Unfortunately, these stored grains are later infested with many pests. Foods stored in bags and bins are very much susceptible to infestation with several pests which can cause extensive post-harvest losses, spoilage, and less demand in markets, causing a huge economic crisis. Hence, successful management of stored grain pests becomes necessary to prevent these from insect pests. Current approaches for their management are one of the promising goals, as it includes preventive practices, monitoring, sanitation, and identification of main pathogens. Different management strategies of all the common stored grain pests viz. grain weevils, grain borers, grain moths, flour moths, mealworms, grain and flour beetles, booklice, mites, and parasites are enlisted here.

Keywords: stored grain insect pests, grain loss management, integrated pest management, economic loss management, pest classification

1. Introduction

Stored grains are heavily damaged by insect pests. These pests cause damage to stored grains resulting in both qualitative and quantitative losses. The main reason behind the occurrence of stored grain pests is the presence of favorable climates for their growth and survival. At various processing stages of grains, i.e., during the process of development and maturation of seeds, processing in threshing yards, during transmission of seeds, or storage large number of insect pests gain access to stored grains. Some pests start damaging the seeds at the ripening stage and continue during storage. Old bags, storage structures, old containers are the major source of infestation [1]. The dispersal and distribution of stored grain pests are caused by the movement of grains from one area to another area either by a passive or active flight of pests as some adult insects possess strong flight. Almost one thousand species are stored grain pests of different stored products all around the world. Undesirable smells and flavor. The majority of stored grain pests belong to two orders, i.e., Coleoptera and Lepidoptera [2].

Stored grain pests possess a serious threat to dried, stored, durable and, perishable agricultural products and non-food derivatives of agricultural products worldwide. Stored grain pests cause serious post-harvest losses, almost 9% in developed countries to almost 20% or more in developing countries [3], besides they also cause contamination of food products by the presence of various live insects, insect products like chemical excretions or silk, dead insects or some other storage structures. Almost 8–10%, i.e., 13 million tons of grains lost due to insects and 100 million tons due to failure to store properly is estimated in stored food products all around the world. Pests such as various insects, pathogens, mites possess serious threats and cause severe damage to grains by producing certain enterotoxins and mycotoxins [4]. Approximately one-third of the world's production, which values almost $100 billion has been destroyed by almost 20,000 species of field and stored grain pests [5]. The majority of stored grain pests belong to the order of Coleoptera and Lepidoptera that accounting for almost 60 and 10% respectively. Of all the stored grain pests [6]. Stored grain pests generally feed on grain, bore into the kernel and then destroy the germ portion, cause heat and then cause deterioration in-stored grain products thus resulting in huge losses mainly due to nutritional depletion and reduction in market value besides cause contamination by their excretory products, that can be extremely hazardous to human health who process and infest the grains so the loss caused by insect pests is not in terms of quantity but mostly in terms of quality. Qualitative loss in stored grain is caused by chemical changes in proteins, carbohydrates, amino acids which negatively affect the nutritional value of grains.

2. Pre- and post-harvest losses by stored grain pests

Grains are generally attacked by several insect pests during all the stages of growth from seedlings to storage [7]. Insect pests possess a major threat to grain production and are also responsible for both direct and indirect losses of grain both in the field as well as in the storage [8]. Mihale et al. estimated that almost 15–100% pre-harvest losses and almost 10–60% post-harvest losses of stored grains are caused by stored grain pests in developing countries [9]. Two major insect groups, i.e., Coleoptera and Lepidoptera are economically important on stored grains. In the case of Lepidoptera, its larva causes the damage while in the case of Coleoptera both larva and adult causes damage.

Weevils and moths are the major stored grain pests that cause huge damage to maize and sorghum [10]. Most important stored grain pests include Angoumois grain moth (*Sitotroga cerealella* (Olivier, 1789) Lepidoptera: Gelechiidae), maize weevil (*Sitophilus zeamais* Motschulsky, 1855 Coleoptera: Curculionidae), the Indian meal moth (Plodia interpunctella (Hubner, 1813) Lepidoptera: Pyralidae), the almond moths, *Ephestia cautella* (Walker) (Pyralidae: Lepidoptera), flour beetles (*Tribolium* spp.), the flat bark beetles *Cryptolestes* spp. (Coleoptera: Laemophloeidae) and the sap beetles *Carpophilus* spp. Stephens, 1830 (Coleoptera: Nitidulidae) [11]. The maize weevil is a major pest mainly found in warm humid areas all around the world. It mainly damages a wide range of cereals and is well established in tropical countries.

Grains such as sorghum and maize are mainly attacked by pests in the field before their harvest. After one week of storage adults of *S. zeamais* were found on all maize portions of the cobs that indicating that cobs are already infested before their harvest. The level of damage to the grains in storage gives an idea about the extent of damage [12]. Maize weevil although commonly found on maize can also attack many cereal grains such as wheat, barley, sorghum, and rice. Although, maize weevil prefers whole grains it has been reported to feed on many processed grain products including pasta and pet food [13].

Almost 10–20% losses have been reported in maize by *S. zeamais* after three months of storage [14]. Thus, millions of tons of maize are lost by stored grain pests due to inefficient storage technologies. More serious damage to maize grains is due to a larger number of adult weevils [15]. It is estimated that almost 63.85% of grain weight losses occurred due to three to six months of storage by stored grain pests.

Pulses are heavily damaged by weevils and beetles in the field and also during storage time [16]. In the case of pulses, the adzuki bean weevil *Callosobruchus chinensis* (Linnaeus, 1758) (Coleoptera: Chrysomelidae) is found to be highly damaging as a stored grain pest. It is estimated that almost 50% of losses are found in important legumes such as chickpea field pea, faba bean by stored grain pests like *C. chinensis* [17]. Bruchids are found to be serious threats to faba bean and chickpea with an extent of damage sometimes reaching 90% after three months of storage.

3. Classification of pests

Stored insect pests are grouped into two types. This grouping is made according to the basis of the feeding ability of the insects in whole or previously damaged grain. They are classified as primary and secondary pests.

3.1 Primary insect pests

Primary insect pests cause damage to the previously undamaged kernel or new grain. Stored grain pests are classified as major and minor pests based on the damage they cause. These insects can be classified as external feeders and internal feeders based on their feeding behavior.

(i) *External feeder*: As the name indicates these pests feed on external or surface parts of the grains such as the outside part of germ and endosperm. These pests either feed on whole seeds or damage the germinal portion of seeds and also feed on those seeds which are already damaged or attacked by other pests or are mechanically broken. These pests are generally visible among the seeds such as rice weevil, pulse beetle, granary weevil, Angoumois moth, etc.

(ii) *Internal feeders*: As the name indicates these pests are usually found inside the seeds. These pests mostly lay eggs inside or on the surface of grains, then spend a part or entire larval and pupal life within the grains and emerge as an adult. These pests cause significant loss of germination that is not detected externally, e.g., rice weevil, pulse beetle, granary weevil, Angoumois moth, etc.

3.2 Secondary insect pests

Secondary feeders. As the name indicates these pests are secondary because these pests attack on already infested crops these generally feed on cut and broken seeds, molds, dead insects, animal wastes, e.g., common mites, cheese mites, etc. Damage caused by these pests results in loss of germination, contamination like ball formation, and webbing besides deterioration of grains. Damage caused by these pests results in fungal activity, moisture migration across the stored grains.

4. Stored grain pests

Some common stored grain pests found all over the world as described below:

4.1 Grain weevils

Weevils or snout beetles (Coleoptera: Curculionidae) have long, elbowed antennae with a special groove on the snout.

*4.1.1 Granary weevil (*Sitophilus granarius *(Linnaeus, 1758) Coleoptera: Curculionidae)*

Distribution: Being cosmopolitan in nature and it is found all around the world.

Host range: This pest mainly feeds voraciously on a large great variety of grains such as oats, wheat, rice, barley, or corn pest.

Bionomics: The granary weevil is the oldest, cosmopolitan, small, brownish or blackish beetle, moderately polished having a long slender snout with a pair of stout mandibles or jaws, and having chewing-type mouthparts [18]. Thorax is well marked with longitudinal punctures and has no wings under its wing covers (**Figure 1**). Larvae or grubs are legless and whitish in color. Adults, as well as larvae, are feeding voraciously on a large great variety of grains. Gravid females lay 200–300 eggs in a small hole in the grain berry with her snout. After oviposition, the hole is covered with a protective gelatinous fluid. Eggs hatched inside holes and white fleshy, legless grubs are formed, which are later transformed into pupae and adults. A short life cycle is seen in summer seasons than in cold seasons.

Damage symptoms: It is one of the most serious pests of grains causing huge damage to the grain. It drastically reduces the crop yields by causing huge damage to harvested stored grains holes are created to the grains that are fed by the pest (**Figure 2**).

*4.1.2 Rice weevil (*Sitophilus oryzae *(Linnaeus, 1763) Coleoptera: Curculionidae)*

Distribution: The rice weevil is highly favored by the hot and humid climate. Being cosmopolitan in nature causes huge economic losses, both larva and adult cause severe damage.

Host range: Crops like paddy, wheat, millet, barley, maize, sorghum dried beans, cotton, nuts, cereals, wheat, corn, flour, pasta, dried flowers, decorative ornaments, stored clothes, dried plants, bread, and other cereals are highly infested by this pest. This pest results in both qualitative and quantitative loss of these crops during their storage.

Bionomics: Rice weevil or Black weevils are small snout beetles, dark brown having 4 distinct patches on the elytra, and prominent spots on the thorax and abdomen. Adults are similar to granary weevils but differ in color, markings, presence of wings beneath wing covers, and thorax with densely pitted with round

Figure 1.
Dorsal view of adult of Sitophilus granarius.

Figure 2.
Damage status of Sitophilus granarius *on wheat.*

punctures (**Figure 3**). Adults are tiny about 2.5 mm long and dark brown in color. Mostly both sexes are alike but in male's rostrum is short and broader. Females lay about 300–400 eggs. Adults are strong fliers and fly from granaries to granaries and to the grain fields for direct infestation. During summer life cycle is very short as compared to winter. Under hot and humid weather eggs take 4–5 days to hatch but under cold conditions eggs take 6–9 days to hatch. The newly hatched larvae bore into the kernel of the grain. Grubs are white in color, curved with a yellow or brown head's and hitting jaws. As grubs emerge from eggs they start feeding on the starchy material of the seeds, till it becomes fully grown and leaves behind only intact pericarp shell which is filled with grass. The grub stage mainly lasts for 19–34 days and then pupates to a non-feeding pupal stage after passing away prepupa for 2–3 days. The pupal stage mainly lasts for almost 1 week and after that adult emerges out of it and starts breeding. This pest completes its life cycle within a month. Most of the severely damaged crops resemble moldy grains.

Damage symptoms: Both larva and adult of this pest are extremely damaging larva enters inside the grain and then starts living and feeding inside the grain due to which irregular holes of about 1.5 mm diameter are produced on the grain. These pests cause extreme damage to stored grains.

*4.1.3 Broad-nosed grain weevil (*Caulophilus oryzae *(Gyllenhal, 1838) Coleoptera: Curculionidae)*

Distribution: Being cosmopolitan in nature, it is found all around the world.

Figure 3.
Dorsal view of adult of Sitophilus oryzae.

Host range: Corn is the main host of this pest. Both fields as well as stored ones are very much susceptible to infestation.

Bionomics: Broad-nosed grain weevil is a small, dark brown with short and broad snout, and similar with granary beetle. It damages soft or damaged seeds and not the dry, hard, and uninjured seeds. These are also strong fliers and can damage the crop fields especially cornfields before the harvesting season. Gravid females lay around 200–300 small whitish eggs inside broken and soft grains. These eggs are hatched in a few days into small, white footless grubs and later into whitish pupae. During summer seasons when environmental conditions are favorable, a very small life cycle can be seen than during harsh cold winter seasons.

Damage symptoms: It damages soft or damaged seeds and not the dry, hard, and uninjured seeds.

*4.1.4 Coffee-bean weevil/nutmeg weevil (*Araecerus fasciculatus *(De Geer, 1775) Coleoptera: Curculionidae)*

Distribution: The coffee-bean weevil (Coleoptera: Brenthidae) is cosmopolitan in nature.

Host range: Its main hosts are dried fruits, coffee, corn, cornstalks, seeds, and seed pods.

Bionomics: The coffee-bean weevil is very active, dark brown in color, with mottled light and dark-brown pubescence, robust beetle. It can be seen in cornfields where they fly here and there. They are usually seen inside soft seeds than hard seeds so they can damage a little to the stored grains gravid females lays eggs inside soft kernels of corn holes.

Damage symptoms: Coffee-Bean weevil can be seen flying in cornfields as well as both larvae and adults inside containers or bins containing grains.

Management of grain weevils:

Freezing is one traditional method in which stored grains are stored in freezing conditions to increase their shelf life free from infestation. Vacuum cleaning is another traditional method in which any stage of any pest can be pulled from any surface with the vacuum cleaners by sucking all of them. Sun drying of grains is also beneficial. Cleaning, damp-proofing, and heating arrangements should be made possible before storing grains in storehouses or godowns. Corn and other husk-bearing crops should be stored in the shuck if the husk is tight, and covers the whole tip, but if all ears with loose, short, broken, damaged, or perforated husks should be shucked and stored separately in clean bins. Placing neem leaves inside grain containers is also recommended. Chemical control can be performed by applying 5% BHC at the rate of 0.15% by weight. Before storage of grains, godowns, containers, and bins should be sprayed with 0.02% Malathion or 0.4% BHC or DDT. Fumigants such as methyl bromide, ethylene dibromide, phostoxin tablets, and HCN are also used for fumigation for 18 h in the closed godowns.

4.2 Grain borers

Grain borers can bore into almost anything such as fabrics, furniture, paper, seed kernels, and seeds.

*4.2.1 Lesser grain borer (*Rhyzopertha dominica *(Fabricius, 1792) Coleoptera: Bostrichidae)*

Distribution: This pest is originated in India but now this pest has spread all around the world. After rice weevil lesser grain borer is considered as second in importance as a destroyer of stored grains.

Figure 4.
Dorsal view of adult of Rhyzopertha dominica.

Host range: Initially, it was mainly found to invest wheat packings but now it's found to be a pest of all cereals. It's mainly found in warmer areas of the world and damages mainly wheat, barley maize, paddy, sorghum, and other crops. It causes its damage by mainly boring into the wood in both larval and adult stages.

Bionomics: Lesser grain borer or Australian wheat weevil are cosmopolitan, small, cylindrical-shaped, dark brown or black in color, roughened surface bodied entities. The head of lesser grain borers are turned down under the thorax and are armed with powerful jaws for cutting and piercing the wood (**Figure 4**). Adults and larvae, both are causing serious damage in warm climates than in cold climatic conditions. Gravid females lay 300–500 eggs loosely or in clusters in the grains. After hatching small, whitish grubs are emerged and voraciously feed on the seeds. Inside grains, these larvae are transformed into pupae and later into adult beetles that came out of the grain through holes.

Damage symptoms: Both larvae and adults of this pest cause serious damage. Highly infested grains become completely hollow inside and only the outer thin shell remains intact. Almost four beetles can be present in bigger grains such as maize. Adults are mostly good fliers so they can easily migrate from one godown to other. Adults produce a considerable amount of frass, spoiling more than what they eat.

*4.2.2 Larger grain borer (*Prostephanus truncatus *(Horn, 1878) Coleoptera: Bostrichidae)*

Distribution: The larger grain borer is originated from India but now this pest is cosmopolitan in nature.

Host range: The main host is corn, wheat, rice, and millet.

Bionomics: A larger grain borer is a small, dark brown, cylindrical, with a smooth polished surface. Both larvae and adults feed on grain kernels and leave dust and thin brown shells. Females lay 2–30 eggs in clusters on kernels which are later transformed into creamy colored C-shaped grubs with a small dark head that is partly retracted into the thorax, having three pairs of small legs. Adults are brown to black cylindrical bodied pests with numerous small pits on wing covers.

Damage symptoms: These are notorious for the emission of sweet, musty odor during an infestation.

Management of grain borers:

The most economical and efficient method of controlling this pest is the prevention of crops. One such preventive method is fumigation. Corn and other husk-bearing crops should be stored in the shuck if the husk is tight, and covers the whole tip, but if all ears with loose, short, broken, damaged, or perforated husks should be shucked and stored separately in clean bins. The application of insecticides is the rapidly controlling method for immediate results.

4.3 Grain moths

Grain moths are found to damage solid, sound, and unbroken stored grains. They reduce kernels as well as grains to powder and shells. Following pests are included in the moth category:

*4.3.1 Angoumois grain moth (*Sitotroga cerealella *Olivier, 1789 Lepidoptera: Gelechiidae)*

Distribution: The name Angoumois is given to this pest because it was first noticed as a pest in the Angoumois province of France in 1973.

Host range: It is mainly found in a warm temperate climate and attacks both stored as well as field grains. It causes huge damage to the grains of paddy, sorghum, bajra, wheat, etc.

Bionomics: The Angoumois grain moth is considered one of the most serious pests or internal feeders in stored grains. It is a small, yellow-brown moth that attacks all cereal grains directly in fields as well as stored ones in all parts of the world. Females lay on or inside grains, around 150–350 eggs, which are initially white and later turned into a reddish color. Eggs are hatched into white caterpillars that are voracious feeders and eat out a channel to the outside of the grain. Larva enters into the grains and starts eating and then turns about and spins a silken web over the opening from which it enters therefore it's difficult to locate the pest. Only larvae are voracious feeders and they feed on kernels. Infestation cannot be assessed in the early stages. Germination is seriously affected after infestation. It is the most devastating pest out of all Lepidopteran storage pests. White larvae are transformed into reddish-brown pupae and later emerge as moths. Adults are usually good fliers. Infestation starts in maturing cereal crops right in the field.

Damage symptoms: The first infestation starts when the grain is in or passing through the milk stage in the field and when only a small percent of grains is infested. By the time grains are threshed or stored infestation increases quickly. When storing the grains infestation of these pests is restricted only to the upper surface. Early infestation is difficult to detect because a hole made by young is so small that it cannot be seen. The first indication of infestation of pests is given by the appearance of moths in the stores and round holes on the grains or sometimes grains get heated up in the bin. Infested grains are hollow insides and filled with excreta or webs of larva leaving a circular opening for moths' emergence. If the pest is breeding in farm godowns, the moth is attracted by instinct to the nearby field in search of maturing grains to lay eggs.

*4.3.2 Wolf moth (*Manduca rustica *(Fabricius, 1775) Lepidoptera: Sphingidae)*

Distribution: The wolf moth is cosmopolitan in distribution.

Host range: It is mainly found in a warm temperate climate and attacks both stored as well as field grains. It causes huge damage to the grains of paddy, sorghum, bajra, wheat, etc.

Bionomics: Wolf moth is a small, creamy white and has a thickly mottled appearance with brown color that distinguished it from the Angoumois grain moth.

Damage symptoms: Infestation can be assessed during early stages with the presence of creamy-colored grubs as well as brown-colored adults.

*4.3.3 Pink cornworm (*Helicoverpa zea *(Boddie, 1850) Lepidoptera: Noctuidea)*

Distribution: The pink cornworm is found all over the world especially in warmer regions.

Host range: They mainly feed on corn seeds, husk, and cob in both fields and storage ones.

Bionomics: Pink corn worm is a small moth having banded fore wings with black, yellow, and brown bands. Hind wings are gray in color, cylindrical, and are edged with long fringes. Females are laying single or occasionally two or three eggs which are white in color. The larvae or caterpillars are pink with pale brown thorax and head.

Damage symptoms: The main indication of its presence in the formation of a large amount of frass that is loosely webbed together and fills the gaps between the kernels.

*4.3.4 Rice moth (*Corcyra cephalonica *(Stainton, 1866) Lepidoptera: Pyralidae)*

Distribution: The rice moth is mainly distributed in Africa, Asia, and Europe. It is one of the most important pests in both India and Pakistan in its larval stage.

Host range: Rice moth is a very serious pest of stored paddy, rice, and other cereals. It is widely distributed in all rice-growing areas. It grows well in humid and warm climates and also infests wheat, sorghum, maize, barley, oilseeds, and sweet products.

Bionomics: Rice moth is pale, grayish-brown in color, and is generally 11–12 mm long. Females are larger than males. Adult life is usually for a week. The Head is provided with tufts of hairs. Almost 200 eggs are laid by females which are small, oval, and are mostly laid on bags, walls. Larvae differ from the larvae of Indian meal moths in having variable color forms such as white, green, and slightly bluish-gray. Larvae feed on rice, biscuits, candies, cocoa, and other kitchen foods. Pupae are pink, elongated, cylindrical with dark spots on the apical side.

Damage symptoms: While feeding, larvae produce dense silken web structures that show their infestation. Besides these pests also pollute the environment with large quantities of frass and silken cocoons, webbing together the grains into large lumps occur.

*4.3.5 Fig-almond moth (*Cadra cautella *(Walker, 1863) Lepidoptera: Pyralidae)*

Distribution: The almond month is also known as fig moth. It is widely spread in the tropics and subtropical areas.

Host range: It causes severe damage to figs, rough rice, dry fruits, wheat, barley, sorghum, oilseeds, etc.

Bionomics: The fig-almond moth (Lepidoptera: Phycitidae) is small, grayish, with transverse spines on the outer wing margin. Nearly 200–250 eggs which are small, oval, whitish in color are deposited by gravid females inside cracks and crevices. Eggs are usually less than 1 mm and hatched around 4 days. Larvae are pinkish-white, living inside the spinning tubes, and later construct silken cocoons for pupation. The pupal period is 7–10 days, which are later emerging into the adult stage, commonly referred to as moths. Moths are generally more abundant during rainy and humid seasons. In certain cases, they only use stored grain pests as their breeding sites and not for feeding purposes.

Damage symptoms: The presence of larvae is a sign of infestation. The larvae make tunnels inside the food grains. They can also block the machinery or mills with clot formation.

Management of grain moths:

Corn and other husk-bearing crops should be stored in the shuck if the husk is tight, and covers the whole tip, but if all ears with loose, short, broken, damaged, or perforated husks should be shucked and stored separately in clean bins.

4.4 Flour moths

The most common and most serious pests of stored grains are flour beetles and flour moths. They eat injured, broken grains, meals, and flour most commonly. Some common flour beetles are:

*4.4.1 Indian meal moth (*Plodia interpunctella *(Hübner, 1813) Lepidoptera: Pyralidae)*

Distribution: Indian meal moth is found all around the world especially in countries that have temperate climates.

Host range: They are mainly damaging dried fruits, nuts, cashew nuts, almonds, etc.

Bionomics: The Indian meal moth is a reddish-brown having peculiar markings on its forewings (**Figure 5**). Females lay 300–400 eggs on food grains either singly or in groups. Eggs, later on, hatched into whitish and sometimes into greenish or pinkish caterpillars that feed voraciously on stored grains, dried fruits, nuts, and many other foodstuffs. Larvae spin a silken cocoon and transform into a light brown pupa from which the moths emerge later on.

Damage symptoms: Full-grown larvae leave behind silken threads wherever they crawl, as well as the presence of greenish or pinkish caterpillars.

*4.4.2 Mediterranean flour moth (*Ephestia kuehniella *Zeller, 1879 Lepidoptera: Pyralidae)*

Distribution: Being cosmopolitan in nature, the Mediterranean flour moth (Lepidoptera: Pyralidae) is found all around the world especially in countries

Figure 5.
Dorsal view of adult of Plodia interpunctella.

that have temperate climates. It mainly prefers warm temperatures for rapid development.

Host range: The larva of the Mediterranean flour moth mainly prefers flour meal, whole grain, and grain residues. This pest shows exception in a way of feeding on cereals rather than feeding on dried fruits.

Bionomics: The Mediterranean flour moth is one of the most serious pests of flour mills, storehouses, granaries, and bran mills that has pale leaden gray fore wings with transverse wavy black markings. Hind wings remain inside fore wings during rest and are white in color (**Figure 6**). They produce such dense webs with flour or meals that can eventually clog mills and the machinery have to shut down thorough cleaning processes. Females lay small whitish eggs in the accumulation of flour, kernel, meal, or waste and crushed grains. Larvae that emerged from these eggs are small, white, or pink in color with a few small black spots on the body. Reddish-brown pupae are formed inside the silk cocoon formed by the full-grown larvae.

Damage symptoms: The infestation can be seen during the production of dense webs with flour or meals inside bins. These can eventually clog mills and the machinery have to shut down for thorough cleaning processes. This species particularly enjoys inhabiting flour mills and bakeries due to the heat, which allows it to breed year-round.

*4.4.3 Meal snout moth (*Pyralis farinalis *(Linnaeus, 1758) Lepidoptera: Pyralidae)*

Distribution: Snout moths (Lepidoptera: Pyralidae), which are also called grass moths or pyralid moths are found throughout the world.

Host range: The meal snout moth is usually found in flour meals and cereals of all kinds.

Bionomics: The meal snout moth is brown, larger than the Indian meal moth with patterned fore wings. Larvae which are black and white when fully grown, usually feed on cereals of all kinds and spin their resting place of peculiar tubes made up of silk and food particles. Fully-grown larvae come out from these tubes, spin silk cocoons, and transform into pupae. During favorable conditions, pupae are transformed into adult moths to start the new generation.

Figure 6.
Dorsal view of adult of Ephestia kuehniella.

Damage symptoms: Silk cocoons inside which pupae are resting are the main indication of this pest. Brown moths are seen flying through the windows of storerooms.

Management of flour moths:

Pest management professionals should be informed as soon as possible to identify the pest properly and to devise the best treatment to control the infestation of food grains. Stored grain containers should be thoroughly inspected for holes, rips, and other larvae or adult presence, before purchasing and after storage. Proper ventilation to prevent moisture build-up, make sure to thoroughly wipe, down, and dust storehouses, cabinets, cupboards, and pantry areas.

4.5 Mealworms

Mealworms (Coleoptera: Tenebrionidae) are dark brown or black or dull pitchy black beetles frequently found in grains especially corn and their larvae are conspicuous and are about an inch in length or as around like an earthworm. Following are some stored grain pests belonging to mealworms:

*4.5.1 Yellow mealworm (*Tenebrio molitor *Linnaeus, 1758 Coleoptera: Tenebrionidae)*

Distribution: The yellow mealworm is the largest of the insect species that attack stored grain and grain products and is cosmopolitan in nature.

Host range: It mainly shows preference to the decaying grains, milled cereals, and usually the foodstuffs which are moist and are going out of conditions. It also feeds on the meal, grain, brand, bread, and dead insects.

Bionomics: Yellow mealworm or darkling beetle is a polished black or dark-brown beetle with finely punctured thorax and with longitudinally striated or grooved fore wings. Females lay about 300–500 eggs, which are bean-shaped, white, and sticky that adhere to food materials with one. Larvae are long, cylindrical, and initially white but later on changed into yellow color. Some larvae are not transformed into adults but instead that they continue their feeding and start molting and undergo hibernation during harsh conditions. Later on, some are transformed into pupae from where adults emerge in the form of beetles.

Damage symptoms: The presence of long, cylindrical, white larvae, usually attached to food materials is the main symptom of an infestation.

*4.5.2 Dark or black mealworm (*Tenebrio obscurus *Fabricius, 1792 Coleoptera: Tenebrionidae)*

Distribution: Being cosmopolitan in nature this pest is found all around the world, especially across Canada.

Host range: The presence of long, cylindrical, white larvae, usually attached to food materials is the main symptom of an infestation.

Bionomics: Dark mealworm resembles a yellow mealworm but differs in being dull pitchy black in contrast to the shiny or polished dark brown or black color. Larvae are long, cylindrical, and initially black but later turned into a blackish color.

Damage symptoms: Larvae are voracious feeders and their presence shows maximum infestation.

Management of mealworms:

Screening and fanning are the best methods for their removal from grain shipments. Good sanitation efforts, inspecting items, and keeping stored grain rooms fully ventilated can help to keep red flour beetles away from entering into food grains. The application of pesticides can be harmful because these pests are found in our food supply. Hence contacting experts for pest solutions at the first sign of an infestation is an effective way to protect food items from red flour beetles. Chemical control can be performed by applying 5% BHC at the rate of 0.15% by weight. Before storage of grains, godowns, containers, and bins should be sprayed with 0.02% Malathion or 0.4% BHC or DDT. Fumigants such as methyl bromide, ethylene dibromide, and HCN are also used for fumigation for 18 h in the closed godowns.

4.6 Grain and flour beetles

The grain and flour beetles are very common stored grain pests feeding on almost all stored grains available throughout the world.

*4.6.1 Cadelle (*Tenebroides mauritanicus *(Linnaeus, 1758) Coleoptera: Tenebrionidae)*

Distribution: Cadelle is present in almost all parts of the world.

Host range: its main host is grain and grain products.

Bionomics: Cadelle is the longest-lived pests, frequently found in granaries, mills, storehouses containing grains, flour, meal, pulses, etc. These are elongated flattened, oblong, black, or blackish beetle that resembles mealworms but, differ from them in size and body texture. These are smaller than mealworms and have loosely joined thorax and abdomen. Larvae are white, fleshy, long, with head, thoracic shield, having two black horny points on the abdomen, and are one of the largest of the stored grain pests. Both larvae and adults are very much destructive and move from grain to grain and devour the embryo.

Damage symptoms: It is a very serious pest of grains and bores inside wood for pupation. The presence of small holes in grains and in wood is the primary indication of its infestation.

*4.6.2 Saw-toothed grain beetle (*Oryzaephilus surinamensis *(Linnaeus, 1758) Coleoptera: Silvanidae)*

Distribution: The saw-toothed grain beetle is cosmopolitan in nature and is found in almost all places of the world.

Host range: The main host includes infesting grains, meals, flour, dried fruits, and many other seeds.

Bionomics: The saw-toothed grain beetle has six saw-like projections on each side of the thorax, with three-segmented antennae. It is cosmopolitan and has a long, slender, dark chocolate brown, much-flattened structure (**Figure 7**). Both larvae and adults are voracious feeders and are very active, hence do not spend their lives within a single grain but crawl as well as infest almost every grain. Larvae are white in color, with black markings, flatform, three pairs of legs, and an abdominal proleg. Larvae construct delicate cocoons by secreting silk-like secretory substances which bind food particles and grains with each other. Inside this cocoon, larvae are transformed into pupae and later into adult beetles.

Damage symptoms: Larvae, as well as cocoon formation, is the primary indication of this pest infestation.

*4.6.3 Square-necked grain beetle (*Cathartus quadricollis *(Guérin-Méneville, 1844) Coleoptera: Silvanidae)*

Distribution: Square-necked grain beetle, was initially found in South America, but now its status is worldwide.

Figure 7.
Dorsal view of adult of Oryzaephilus surinamensis.

Host range: Being a stored grain pest it generally attacks the stored products such as cereals, grains, etc.

Bionomics: The square-necked grain beetle is a flattened, oblong, polished, reddish-brown, with thorax almost square-shaped. These are some of the most common beetles found in both cornfields as well as granaries. Usually, after three weeks eggs get converted into larva, the larva undergoes molting five times to become a pupa, pupa then again gets converted into adult [19]. Eggs are mainly oval in shape and are opaque white in color and are less than a millimeter long, 4 days eggs hatch into larva and the larva are the main predators [20]. The pupa is generally darker in color and then the pupa gradually transformed into adults.

Damage symptoms: Larval presence is the main symptom of infestation by this pest.

*4.6.4 Foreign grain beetle (*Ahasverus advena *(Waltl, 1834) Coleoptera: Silvanidae)*

Distribution: Foreign grain beetle mainly occurs in the tropics and sub-tropics. This pest can complete the development at temperatures between 20°C and 35°C.

Host range: This pest mainly attacks cereals, grains, oilseeds, spices, and dried fruits.

Bionomics: The foreign grain beetle is a small, cosmopolitan, reddish-brown beetle. It spends its most time on damp and moldy grains and less on clean grains. This pest mainly measures about 2 mm in length. It can be easily differentiated by the slight projections on each front corner of the pronotum and the antennae are club-shaped. The larvae are mainly cream-colored, worm-like, and almost 3 mm long before they start pupation into darker adults. Males and females are alike in form both as larva as well as adults.

Damage symptoms: Poor storage conditions and spoiled food is mainly indicated by the presence of the beetle.

*4.6.5 Mexican bean beetle (*Epilachna varivestis *Mulsant, 1850 (Coleoptera: Coccinellidae)*

Distribution: The Mexican bean beetle is mainly found in Mexico and the Eastern United States. It is largely found in wet and highly irrigated areas especially in the west of the Rocky Mountains. It cannot tolerate extremely dry conditions.

Host range: This pest is mainly feeding on flower, leaf, or pod tissues on beans and other legumes. The pest is generally found in different varieties of bean plants including thicket bean, cowpea, common bean, soybean. Besides it also feeds on other legumes such as alfalfa and many others.

Bionomics: The Mexican grain beetle is a deep brown, highly polished, long antennae, and usually breeds in grain and grain products. The adult is mainly oval shaped almost 6–7 mm in length it bears eight black spots on its elytron. The Color of the adult is highly variable ranges from bright red to rusty brown to golden yellow. Almost 1.3 mm long eggs are generally yellow and are glued in clusters on the underside of the leaves. The larvae are yellow in color. Almost 1.6 mm long when they first emerge but they grow almost a centimeter long before pupation.

Damage symptoms: Places wet and highly irrigated are the main infested regions by this pest.

*4.6.6 Siamese grain beetle (*Lophocateres pusillus *(Klug, 1832) Coleoptera: Trogositidae)*

Distribution: Siamese grain beetle is found in all regions where paddy and different kinds of cereals are cultivated.

Host range: The main host is rice and cereals.

Bionomics: The Siamese grain beetle is reddish-brown, flattened, elongate, long beetle with a flattened margin of the thorax and wing cover. These are making holes in the rice and hence damage its quality as well as quantity.

Damage symptoms: Adults are seen actively walking on and around the food grains.

*4.6.7 Flat grain beetle (*Cryptolestes pusillus *(Schoenherr, 1817) Coleoptera: Silvanidae)*

Distribution: Flat grain beetle is commonly found all over the world.

Host range: Its main hosts are corn and wheat grains.

Bionomics: The flat grain beetle is cosmopolitan, small, oblong, flat, reddish-brown beetle, with long antennae. Larvae woven cocoons by secreting sticky secretions, adhering damaged grains with each other for pupal development.

Damage symptoms: Flat grain beetles have the property of secreting sticky secretions.

*4.6.8 Confused flour beetle (*Tribolium confusum *Jaqcquelin du Val, 1868 Coleoptera: Tenebrionidae)*

Distribution: The confused flour beetle, is cosmopolitan, native to Africa, and commonly found in cooler places.

Host range: It is commonly found in flour mills, granaries, storehouses, wheat fields, dried flowers, seeds, or dried museum specimens [21].

Bionomics: The confused flour beetle is reddish-brown, shiny, long, oval, flattened, having four segmented antennae, with head and upper parts of thorax densely covered with small punctures and with ridges on wing covers (**Figure 8**). Eggs are small white in color, laid by gravid females inside boxes, barrels, and other food containers. These sticky secretions help them to adhere to the flour as well as with the walls of containers. Eggs are transformed into long, worm-like larvae which are cylindrical and wiry in appearance. The pupal stage is small, initially white and later yellow and brown, where adult beetles emerge shortly.

Figure 8.
Dorsal view of adult of Tribolium confusum.

Damage symptoms: Being secondary pests they do not directly attack the grain bur when the grain is already infested, they show their effect. These pests generally give an unpleasant odor and also due to their presence, the growth of mold is encouraged.

*4.6.9 Rust-red flour beetle or red flour beetle (*Tribolium castaneum *(Herbst, 1797) Coleoptera: Tenebrionidae)*

Distribution: Being cosmopolitan in nature red flour beetle, is mainly found all around the world.

Host range: It is a serious pest of cereal products, including grain, flour, porridge oats, and rice bran. Other products which may be attacked are oilseed, oil cake, nuts, dried fruit, spices, chocolate, and even bones of animals.

Bionomics: The rust-red flour beetle is cosmopolitan, shiny, reddish-brown, has antennae enlarged at the tip with the three-segmented club, head margins are continuous and not expanded and notched at the eyes (**Figure 9**). They are notorious for causing bad smells and tastes imparted to the food materials they infest. Their main hosts are maize, wheat, and other mills and granaries.

Damage symptoms: When present in abundance, this beetle makes the flour prone to molding and also turns the products into a gray color.

*4.6.10 Long-headed flour beetle (*Latheticus oryzae *Waterhouse, 1880 Coleoptera: Tenebrionidae)*

Distribution: This pest was first reported from India in Kolkata in 1880. Later it was reported from many other countries.

Host range: This mainly attacks broken grains, wheat, rice, corn, flour, barley, and many other granaries, grocery stores, and mills.

Bionomics: The long-headed flour beetle is cosmopolitan, pale yellow, slender, flattened beetle, with slightly bulged antennae, and the presence of canthus behind each eye. It has been reported from wheat, rice, corn, flour, barley, and many other granaries, grocery stores, and mills. It is mainly associated with *T. castaneum*, and its behavior and life cycle are very similar to flour beetle. Eggs are mostly smooth and translucent in color. Grubs are generally white in color with dark eyes. Larval body is covered by pale-colored hairs. The life cycle of the pest is completed in 25–39 days.

Figure 9.
Dorsal view of adult of Tribolium castaneum.

Damage symptoms: Milled products are fed by both grubs and adults as well. Occurs as secondary infestation in stored sorghum, wheat, etc.

*4.6.11 Slender-horned flour beetle (*Gnathocerus maxillosus *(Fabricius, 1801) Coleoptera: Tenebrionidae)*

Distribution: The slender-horned flour beetle is cosmopolitan in nature.

Host range: It is most commonly found in flour, meal, and a variety of grains.

Bionomics: The slender-horned flour beetle is cosmopolitan, flat, brown, with a pair of incurved horns on its mandibles of the male partner.

Damage symptoms: Being secondary pests they do not directly attack the grain bur when the grain is already infested, they show their effect. These pests generally give an unpleasant odor and also due to their presence mold growth is encouraged.

*4.6.12 Broad-horned flour beetle (*Gnatocerus cornutus *(Fabricius, 1798) Coleoptera: Tenebrionidae)*

Distribution: Broad-horned flour beetle is especially found in Canada and is distributed all over the world.

Host range: It is commonly found in granaries, mills, and many other stored grains.

Bionomics: The broad-horned flour beetle is slender, elongated beetle, with mandibular broad and stout horns in males. After mating female lays eggs either singly or in batches within the food source. From the eggs larva hatch and then they start feeding and gets converted into an adult again. This species primarily feeds dead insects besides feeding on protein sources. Adults show sexual dimorphism. Horns are absent in females but they are present in males.

Damage symptoms: The presence of dead insects inside grains is the prime indication of pest attack.

*4.6.13 Small-eyed flour beetle (*Palorus ratzeburgi *(Wissmann, 1848) Coleoptera: Tenebrionidae)*

Distribution: Small-eyed flour beetle is cosmopolitan in nature.

Host range: This pest generally prefers milled wheat, stored grain, oat products, flour mills. It also feeds on plant and dried animal products such as grain and cereal products. It is mainly a secondary pest, also feeds on fungus, and acts as a scavenger.

Bionomics: The small-eyed flour beetle is one of the smallest, flat, shiny, reddish-brown, oblong flour beetles. It is cosmopolitan and is found in ground products where they feed and breed. Eggs laid by adult females are generally sticky due to which they become coated with flour or grain dust. The larva is highly active and moves freely among the foodstuffs. Adult ate usually 2–3 mm in length and are small reddish-brown in color. The larva is generally cylindrical in shape. This pest is considered as one of the smallest flour beetles and they are generally differentiated by the presence of the eye entirely and not incised by the margin of the head.

Damage symptoms: Although, damage cannot be assessed clearly, however, quality of stored grains is highly affected.

4.6.14 Tobacco beetle or Cigarette beetle (Lasioderma serricorne *(Fabricius, 1792) Coleoptera: Anobiidae)*

Distribution: Although, the Cigarette beetle is a cosmopolitan pest it generally prefers to be in warmer environmental conditions.

Host range: It feeds on a wide range of food materials from spices, chocolate, cocoa, and tobacco leaves. The other hosts are paprika, dry dog food, beans, dried fruits, biscuits, grains, peanuts, rice, and vegetables.

Bionomics: The cigarette beetles are light brown, oval-shaped beetle, having serrated antennae, strong humped appearance on the head and thorax. Egg-laying occurs either in folds or in crevices of food material. Eggs are mostly oval in shape and white in color but become opaque before hatching. Almost 100–110 eggs are laid by females that hatch in 5–6 days. The larval stage lasts for 20–25 days followed by the pupal stage. The larvae are smaller than the adults and are worm-like hence known as cigarette beetles, and tobacco beetle because of residing inside tobacco. The pest cause damage by making little gallerias. After 25–39 days of larval life, it makes smooth-lined cells under which the pest rests. The newly formed pupa is glossy white but gradually changes to reddish-brown in color after a few days. Females are mainly larger in size than males.

Damage symptoms: Both adults and grubs of this pest enter into the tobacco products viz., cigarettes, cheroots, and chewing tobacco. A typical symptom of attack of this pest is the presence of circular pin-sized boreholes on the processed tobacco. Besides this pest also damages cocoa, wheat, cotton seeds, etc.

4.6.15 Drug-store beetle (Stegobium paniceum *(Linnaeus, 1758) Coleoptera: Anobiidae)*

Distribution: Drug-store beetle is generally found in tropical, subtropical, and temperate regions.

Host range: It mainly infests turmeric, ginger, pepper, coriander seeds, cumin, seeds. Adults and grubs mainly attack the grains and seeds. It is frequently seen in drug stores where it can feed and breed. Stored grain foods, seeds, flours are the common hosts of this beetle.

Bionomics: The drug-store beetle is an elongate, cylindrical, light brown, with its body covered with fine silky hair. Females lay almost 50–80 eggs inside grains and other stored substances that are later transformed into small white grubs after 8–10 days of hatching. The larval period lasts for 4–5 weeks which is followed by a pupal period of 6–10 days. Larvae then woven cocoons resulted in pupae that gave rise to adult beetles. Adults are pale brown and short-lived.

Damage symptoms: Damage caused by the pest is indicated by the presence of circular pinhead-sized boreholes on turmeric, coriander, dry vegetables, and animal matter.

*4.6.16 Black carpet beetle (*Attagenus unicolor *(Brahm, 1791) Coleoptera: Dermestidae)*

Distribution: The Black carpet beetle is cosmopolitan in distribution.

Host range: Larva of this pest is a voracious feeder and it mainly feeds on natural fibers, or furniture or carpets, or even clothes.

Bionomics: The black carpet beetle is small, with its head and thorax black colored, and wings either black or reddish-brown or golden brown, clothed with short hairs. Legs and antennae are yellowish in color. Mostly egg-laying occurs on a food source or sometimes females lay eggs in dark undisturbed areas where the larva starts feeding on carpets or clothes. Almost 5–20 days are taken by the eggs to hatch depending on external conditions like humidity, temperature. The larva is mostly 1 mm long when they hatch from the eggs they will grow faster if food sources are abundant. Almost 10–15 molting a larva is usually taken by the larva to undergo molting. The larval phase is the longest phase and then the larva converts into a pupa. Initially, the pupa is creamy colors but they quickly then turn first yellow and then dark in color. Within 8–20 days the pupa transforms into adult beetles. Adult lives only to mate and then lay eggs and finally die. Since in the end, they are black colored that's why they are commonly called black carpet beetle.

Damage symptoms: It is one of the serious pests of grains. Its larval stage is highly damaging, as they are voracious feeders and their presence is the indication of an infestation.

*4.6.17 Larger cabinet beetle (*Trogoderma granarium *Everts, 1898 Coleoptera: Dermestidae)*

Distribution: Larger cabinet beetle is mainly found in tropics and subtropics. It prefers to live in humid and high-temperature areas.

Host range: Being an external feeder it is only found on the surface of the grain. It is the main pest of wheat but, can also destroy jowar, rice, maize, sorghum, oilseeds, and pulses. These are also commonly found in beans, pumpkin seeds, gourd seeds, and many other grains.

Bionomics: The larger cabinet beetle is small, egg-shaped, with a black body mottled with reddish-brown, presence of hairs having gray and light brown color. The adult is usually oval in shape with gray and pale brown markings. The Head is primarily hidden under the hood like pronotum. Almost 100–120 eggs are laid by females after breeding. It takes 5–6 days for females to lay eggs after breeding. The larva is mainly brown in color, the whole body is covered by bundles of long, reddish-brown movable and erectile hair present on the posterior segments which form a sort of tail in the posterior end. First instar larva mainly feeds on broken grains and debris. The larval period persists up to 20–25 days, whereas the pupal period persists for 4–8 days. This pest is highly resistant to starvation. Although, this pest damages whole grain it primarily prefers germ portion due to which viability of seeds is lost long before any quantitative damage occurred. High infestation results in a reduction of whole grains to mere Fras. The larval stage is the devastating stage. Adults are non-feeders.

Damage symptoms: Only larvae are voracious feeders and feed on grain kernels. Holes of almost 1 mm diameter are seen on the grains. This pest imparts an extremely unhealthy appearance and unpleasant smell. Mostly the upper layer of the heap is severely damaged.

*4.6.18 Small cabinet beetle (*Trogoderma *sp. Coleoptera: Dermestidae)*

Distribution: Small cabinet beetles are restricted to warmer regions as well as tropical regions.

Host range: These are usually found in flour mills, granaries, and storehouses.

Bionomics: The small cabinet beetle differs from the larger cabinet beetle in size and color. These are usually small and black with yellowish-white scales on the body. Eggs, as well as larvae, are found inside piercing and broken grains.

Damage symptoms: Grains on keen inspection can be seen soft inside and broken because of its presence.

*4.6.19 Museum beetle (*Anthrenus museorum *(Linnaeus, 1761) Coleoptera: Dermestidae)*

Distribution: Museum beetle is mainly found in Palearctic areas including Europe, the Nearctic, and the Near East.

Host range: This pest mainly prefers flour, cheese, or cocoa.

Bionomics: The museum beetle is black, having yellowish and whitish scales on its body. Almost 50 eggs are oviposited inside grains and the larvae are mainly 4.5 mm in length and bear active hairs, hence commonly referred to as a hairy grub. The dorsal surface of the prothorax is brown in color. It possesses 3 pairs of long antennae at its rear end. The adult is about 2–4 mm in length. It is round in shape. After mating females lay eggs in carpets, flooring, to hide the eggs and to provide food supply to the larva. They are found in stored grain containers but their damage-causing status is very poorly reported.

Damage symptoms: As far as the damage is concerned larva are highly damaging and they mainly destroy all forms of dry grains and flour.

4.6.20 Two-branded fungus beetle

Distribution: The two-branded fungus beetle (Coleoptera: Endomychidae) is cosmopolitan in distribution.

Host range: They mainly feed on fungus and molds and are also frequently found in mills, granaries, storehouses, etc.

Bionomics: The two-branded fungus beetle is small, cosmopolitan, reddish-brown in color with two broad black bands across the wings. Although, feeding on fungi and molds but are also frequently found in mills, granaries, storehouses, etc. Eggs are commonly laid inside infested or damaged grains and the larvae are voracious feeders and spoil grains and cereals.

Damage symptoms: The larvae are voracious feeders and spoil grains and cereals, reducing their quality and quantity status.

*4.6.21 Black fungus beetle (*Alphitobius laevigatus *(Fabricius, 1781) Coleoptera: Tenebrionidae)*

Distribution: Black fungus beetle is cosmopolitan in nature, found all around the world.

Host range: This pest feeds on a large variety of stored products and is also a fungal feeder. It is a secondary pest which means it enhances the damage caused by primary pests.

Bionomics: The black fungus beetle is small, with a black or reddish-brown colored body. They frequently feed and breed in damp moldy grains. Adults are almost 5–7 mm. Based on the lateral view of eyes; adults can be easily distinguished from lesser mealworms. Larval is cylinder-shaped and is yellowish-brown in color. The larva is active and moves quickly towards the food sources.

Damage symptoms: Being a secondary pest it does not directly attack the grains but causes damage in presence of the primary pest. Its presence indicates poor storage and poor sanitation conditions.

*4.6.22 Corn sap beetle (*Carpophilus dimidiatus *(Fabricius, 1792) Coleoptera: Nitidulidae)*

Distribution: Corn sap beetle is cosmopolitan in distribution and mainly originated in the USA.

Host range: It feeds on rotten and decaying fruits and vegetables, corn, and solid grains.

Bionomics: The corn sap beetle is small, oblong or ovoid, dark-brown beetle with short and truncate fore wings with the uncovered abdominal tip.

Damage symptoms: These pests are notorious for the emission of foul smells.

*4.6.23 Pulse beetle (*C. chinensis *(Linnaeus, 1758) Coleoptera: Chrysomelidae)*

Distribution: Pulse beetle is distributed throughout the temperate regions of the world.

Host range: C. chinensis, is a frequent pest of all pulses, beans and grams.

Bionomics: Gravid females lay single eggs, glued to the surface of pods or grains. Eggs are translucent, orange, or cream colored, changing grayish to white later. Eggs hatch into fleshy, curved, creamy white larvae with black mouth parts. Pupae take place inside seed coats in pupal cells. Adults are short, active, brownish-gray, with characteristic spots near the middle of the dorsal side. Adults are not feeding on storage products and are short-lived.

Damage symptoms: Adult are seen emerging and wandering over the surface of the grain, and making exit holes. Grubs are responsible for the formation of cavities in seed kernels.

Management of grain and flour beetles:

Pest management professionals should be informed as soon as possible to identify the pest properly and to devise the best treatment to control the infestation of food grains. Stored grain containers should be thoroughly inspected for holes, rips, and other larvae or adult presence, before purchasing and after storage. Proper ventilation to prevent moisture build-up, make sure to thoroughly wipe, down, and dust storehouses, cabinets, cupboards, and pantry areas.

Infested products with cigarette beetles should be discarded as soon as possible. Stored grain products should be kept in glass sealed containers, plastic containers instead of their original packing. Cleaning and wiping down those areas commonly occupied with food debris.

Corn and other husk-bearing crops should be stored in the shuck if the husk is tight, and covers the whole tip, but if all ears with loose, short, broken, damaged, or perforated husks should be shucked and stored separately in clean bins.

Good sanitation efforts, inspecting items, and keeping stored grain rooms fully ventilated can help to keep red flour beetles away from entering into food grains. The application of pesticides can be harmful because these pests are found in our food supply. Hence contacting experts for pest solutions at the first sign of an infestation is an effective way to protect food items from red flour beetles. Chemical control can be performed by applying carbamates, malathion, organophosphates, organochlorines, etc. These pesticides are used against many stored grain pests. New practices such as ozonation and organic pesticides have ensured grain preservation without quality loss and residue accumulation. Nitric oxide (NO), a newly discovered fumigant, has shown a great potential to control stored grain pests and has been described as a substitute for Methyl bromide.

4.7 Booklice (Psocoptera)

Distribution: Booklice (Psocoptera: Liposcelididae) is cosmopolitan in nature they are found all across the world, and mainly found in old books where they feed on the paste that is used in binding. These are very frequently found in grains, granaries, cupboards, and other solid food substances.

Host range: This pest generally feeds upon algae, fungi, lichen, organic detritus in nature, but they are mostly considered as stored grain pests as they feed on grains, bookbinding, etc.

Bionomics: The booklice or psocids are small, pale, louse-like, soft-bodied insects, with long slender antennae. Eggs of the pest are mainly laid in crevices or on foliage. Nymphs undergo molt for 6 times to reach adulthood. Length of booklice ranges from 1 to 2 mm.

Damage symptoms: Besides damaging books, they also sometimes infest food storage areas, where they feed on dry, starchy materials. Although, some psocids feed on starchy household products, the majority of psocids are woodland insects with little to no contact with humans, therefore they are of little economic importance. Booklice are scavengers and usually do not bite humans.

Management of booklice:

Cleanliness is one the most successful solution against the attack of booklice. Old books should be placed in cooler conditions, free from moisture and high temperature. Naphthalene balls should be placed on shelves and cupboards. Neem leaves should be placed inside bins or containers, containing food grains and other products.

4.8 Cereal mites (*Acarus siro* Linnaeus, 1758 Sarcoptiformes: Acaridae)

Distribution: Mites are microscopic and are cosmopolitan in distribution.

Host range: They mainly attack stored grain pests and rapidly increase their number within a short duration. Almost all plant and animal materials are directly or indirectly affected by these mites.

Bionomics: Mites are soft-bodied creatures, pale-colored, microscopic entities. They mainly attack stored grain pests and rapidly increase their number within a short duration. They can infest the crops either directly or indirectly. Mites shed their skin and dead bodies accumulate in fluffy bright brown masses beneath the sacks of food grain.

Damage symptoms: Decolouration or fading is the prime symptom of any mite attack.

Management of mites:

Biological control is one of the eco-friendly controlling strategies in which some predatory mites usually attack these grain mites and kill them. Manual method: Screening and fanning of grains will reduce their population and check the infestation level.

5. Management of stored grain pests

Insects are notorious to cause enormous damage to grains, pulses, and many other substances either directly or indirectly by consuming the seeds or seed products or through the accretion of exuviae, cadavers, and webbing. Hence making the stored products unfit and unhygienic for human consumption due to the accumulation of insect detritus [22]. Stored grain pests can infest almost all grains stored inside bins or containers as well as outside the fields and cause extensive post-harvest damage and pose a great threat to the economy. Once an infestation happens, a suitable environment is created for the attraction of other invasive insects for further loss. The most consumed and the most common stored food products are pulses and food grains in the tropical and sub-tropical regions of the world. In villages, about 70% of grains produced are stored in traditional objects such as earthen pots, steel drums, granaries, silos, gunny bags, baskets, and wooden buckets [23], such types of storage methods may often lead to loss of food grains and pulses [24]. Controlling strategy without synthetic pesticides requires an Integrated Pest Management (IPM) approach. The IPM approach is not based on a single component instead it is based on various components for the efficient management of insect pests. These components are described here.

5.1 Sampling

Sampling or pest monitoring is an important component of the IPM approach with which one can know the nature of pests in full detail so that suitable management tactics should be made accordingly. With the help of sampling, one can show the status of a pest, whether the population is below or exceeds the economic thresh hold level, and accordingly, physical, biological, or chemical approaches can be recommended. Some methods used during sampling processes of stored grain pests are:

5.1.1 Sequential sampling method

Sampling should be performed frequently after fixed intervals for best observations, and to gather information about population changes from time to time. For example, those stored grain pests stored above 20°C should be visited after a gap of 25–30 days. Grains held below 20°C permits sampling intervals to be longer than 25–30 days.

5.1.2 Population density estimation method

i. *Absolute estimation*: In this method number of insects per kilogram of grain or the number of moths per square meter are estimated.

ii. *Indirect estimation*: Here pests are marked with a specific dye and then recaptured after releasing into the stored grains, hence commonly referred to as mark-release-recapture methods. It can be easily performed with the help of suitably designed traps with baits.

iii. *Relative estimation*: This method can be done by counting all the insects caught in a sticky trap, food baited trap or perforated probe trap.

5.1.3 Trapping method

Trapping is a convenient approach in small as well as in the larger volumes of grain containing granaries and fields as well. Sticky traps, food-baited traps, pheromone traps, or perforated probe traps are used for monitoring processes.

5.2 Preventive measures

Infestation can be entirely prevented when some precautionary measures should be taken such as when harvesting crops should be as soon as ripe, dry, and then placed in clean, and hygienic deep bins for long storage. Newly harvested small grains are very much susceptible to infestation if stored unthrashed for longer times. Fresh and clean grains should never be stored in uncleaned, old bins and granaries containing waste grains, until they have been thoroughly cleaned, freed from the accumulation of waste materials and other substances harboring grain pests. The best storage places are solid, steel, concrete bins or containers for infestation-free and for longer storage. Traveling bags, bags used for transportation of grains, and any other products should be kept far away from the places where grains are stored.

5.3 Traditional practices

From time-to-time man has continuously developed various conventional methods to protect stored food grains from insect damage. Use of bamboo, wooden plank, straw, mud, bricks, cow dung, leaves of many plants, etc. is used by farmers to protect the quality as well as the number of stored foods until for further consumption [25]. One of the most common methods used by farmers was the use of plant parts or plant extracts as natural insecticides and repellents. During the 1850s, plants such as *Nicotiana tabacum*, *Derris elliptica*, *Lonchocarpus* spp., *Juglans regia*, *Azadirachta indica*, and *Chrysanthemum cineraria* folium was used for the plant extracts such as nicotine, derris dust, rotenone, Juglans, Azadirachtin, and pyrethrum respectively for controlling pests naturally [26]. The discovery of DDT by Paul Muller marked the advent of a new synthetic pesticide era since 1939.

5.4 Organic approach

The list of all usable, as well as prohibited controlling methods, are permissible in the national organic program (NOP). All the generic materials are enlisted under the national list of allowed and prohibited substances (NLAPS). It is mentioned in this that organic control should be the top priority, although synthetic insecticides can also be used upon specific approval. Certification to every producer, controller, processor, and handler is mandatory for authorized permissible processes. To reduce the infestation of stored grain pests, we should not make ourselves victims of pesticides. For this wearing the appropriate protective clothing and equipment during pest control to avoid contact to eyes, lungs, skin, and nose. Some control materials allowed in organic stored grains are:

Bacillus thuringiensis: This bacterium is used to control and prevent pests especially the larvae of Indian-meal moth. *B. thuringiensis* damages the digestive tract of caterpillars and lastly kills them.

Pyrethrum: Botanicals based on pyrethrin obtained from the flowers of *Chrysanthemum cinerariifolium* are primarily an insecticide that penetrates rapidly inside insect coverings, especially moths and larvae [27]. Empty containers should be treated before they are filled with grains for best results. Pyrethrum is an insecticide that is now universally accepted and is used to reduce pest damage in both tropical and temperate climatic conditions [28].

Diatomaceous earth: Aquatic organisms commonly referred to as diatoms have their skeletal system made of silica. The fossilized forms, having sharp edges of these diatoms are commonly referred to as diatomaceous earth. The sharp edges of diatomaceous earth can cut the pest's cuticle, resulting in death by injury and dehydration.

Grain surface protectant: Cleanliness is an essential factor to lower the damage rate. Containers and bins are filled only to the height of sidewalls, floors, and ceilings, and then cleaned through the fan system. Topdressing or simply capping the stored grains will act as a protective barrier from migrating insects into the bin.

Grain rescue: Infested grains should be treated initially with some treatments such as appropriate cooling and warming before being used for food to humans or any other animal.

Detech and methyl eugenol: These are promising treatments for the control of stored grain pests such as *S. granarius*, *S. zeamais*, (Coleoptera: Curculionidae),

Rhyzopertha dominica (Coleoptera: Bostrychidae), *Tribolium confusum* (Coleoptera: Tenebrionidae). ME is a benzene-derived component, potential, and effective plant-derived synthetic chemical insecticide, and has a high knockdown effect because of the presence of more methoxy groups in it [29]. A synergistic effect of the combination of Diatomaceous Earth and Methyl Eugenol on *R. dominica*, *T. confusum*, *S granaries*, and *S. zeamais* has been reported by Erturk in 2021) [30].

5.5 Physical methods

Once the stored grains are infested, some physical methods used for the management of the stored grain pests are:

Physical exclusion: Fine perforated floors are made for the collection of dusty fines at the bottom that are susceptible to insect infestation.

Grain distribution: Grains inside granaries as well as inside bins and containers should be properly leveled. Improper leveling can create room for insect infestation and mold development due to the accumulation of moisture into the peaked-grained mass. To prevent the stored grains, removing grains from the old bins and redistributing them to other containers are very helpful.

Temperature: Based on the nature of pests, the temperature can be set either at low or high degrees. As some pests like moist and cool places and some like hot and humid regions. Most pests require temperatures above 60–70°F to reach damaging populations. Hence maintaining a cool temperature can reduce the excess loss. In certain situations, maintenance of −4°C to 0°C can kill many stored grain pests. *T. castaneum* and *Oryzaephilus mercator* are highly susceptible to cold, whereas *Trgoderma* spp., *Plodia interpunctella*, and *Ephestia* spp. are cold-tolerant species. Maintenance of very high temperatures can also be recommended but it has certain drawbacks such as it can crack, harden, and make brittle grains inside bins.

Hermetic sealing: To maintain a very low oxygen level inside stored grain containers this method is used. Low oxygen level causes suffocation to the pests and hence has insecticidal property.

Aeration: Air flown at the rate of 0.1–0.5 cubic feet per minute per bushel are used to cool stored grains. This low-volume airflow is an important component of the management of the stored grain pests. Grains remain uniform and to some extent in dry conditions as some grains are susceptible to pest attack in moist climates.

Oxygen saturation: Insects perform aerobic respiration for their survival. Maintenance of low O_2 atmosphere is blown at the base of the containers, bins, and other stored chambers, forcing out the existing O_2 rich atmosphere is a convenient method for infestation control.

Sanitation: All bins, containers, granaries, and other stored places should be cleaned using shovels, brooms, vacuum cleaners to clear old grains, dust, spider web, and fines from all cracks and crevices, windows, doors, vents, fans, elevators, and floor. Even a small old grain or fines left in any place where new grains are to be stored can harbor insects that can infest the whole grain. A suitable dryer should be used to remove the moisture from bins. To improve storability, especially in the case of wet, damaged, or immature grains, grain cleaners can be used frequently. Some grain cleaners are:

i. Gravity screens, with which grains are passed over a screen during handling.

ii. Rotary screens, are very effective cleaners that tumble and separate fines from grains.

iii. Perforated auger, is used to separate fines when the grain is conveyed over the auger.

iv. Aspirator pre-cleaner, removes all those materials which are lighter than grains such as dust, husk, awn, etc. by flowing air through it.

5.6 Conventional methods

Since the discovery of DDT, the use of synthetic insecticides was established as one of the most reliable and successful controlling agents worldwide [31]. No, any method is so rapid in action as synthetic chemicals are, hence farmers are indiscriminately using them without keeping any precautionary measures. Indiscriminate usage of synthetic insecticides has been characterized by several negative impacts such as resistance, toxicity, ozone depletion, adulteration, erratic supplies, and unavailability at critical periods [32].

Fumigation: Some most common fumigants used for treating stored grains in bulk are carbon tetrachloride, carbon disulphide, methyl bromide, phosphine, and hydrocyanic acid. However, methyl bromide is listed as an ozone-depleting compound in 1993 and has been phased out as the Montreal protocol [33]. Instead of methyl bromide, phosphine is used to protect the food grains as well as other products such as spices, cocoa beans, dried fruits, nuts, and even fresh fruits [34]. Fumigation is one of the convenient methods and the fumigants are heavier than the air and when applied on the top of the gas-tight bin of stored grains will penetrate down through the grains, killing the pests of any stage and without any harm to the grains. Fumigation should be done under a precautionary setup as these gases are highly inflammable and will explode if a fire is brought near them.

The insecticides should not be sprayed directly on food grains. Instead, treat the walls, dunnage materials, and ceilings of empty godown with malathion 50 EC 10 ml/L. Treatment of alleyways and gangways should be done with malathion 50 EC 10 ml/L. Spraying of malathion 50 EC 10 ml/L with @ 3 L of spray fluid/100 m^2 over the bags and other containers. In the case of flying insects and insects on surfaces, cracks, and crevices, a spray of pyrethrum seems good in action. Before storage, seed protectants like pyrethrum dust, carbaryl dust to mix with grains should be used. Ampoules of EDB should be used at 3 ml/quintal for wheat and pulses and 5 ml/quintal for rice and paddy. One of the most crucial fumigants for the control of stored grain pests is Phosphine. However, it may raise human safety concerns as phosphine is a poisonous gas and is known to be adsorbed in grains during fumigation. Nanoencapsulation of 25 kDa cysteine protease obtained from *Albizia procera* (ApCp) could be a promising ecofriendly tool of insect pest control.

5.7 Biological control of stored grain pests

Biological control includes the use of some predatory insects or microbes to control pests. Some beneficial insects such as hymenopterous parasites are attacking and killing many stored grain pests such as weevils, rusty grain beetle, maize weevil, confused flour beetle, lesser grain borer, Angoumois grain beetle, sawtooth grain beetle, and grain moths. Parasites are killing a large number of grain pests, but are not providing complete protection as the grains themselves have become very badly damaged. Small black wasp-like insects (*Seenopinus fenestral*) are also feeding and rearing on many stored grain pests and help to decrease their infestation. Larvae of a window-pane fly are thread-like white worm that does not harm grain but acts as predacious upon many grain pest larvae. Another parasitoid wasp *Theocolax elegans*, attacks primary grain pests whose immature stages are grown inside seed kernels, including the lesser grain borer, weevils, the drug store beetle, cowpea weevils, and the Angoumois grain moth [35]. In Europe, *Trichogramma* spp. has been used against moths in groundnut, wheat, bakeries as well as in warehouses and retail shops [36]. *Dinarmus* spp. is a larval and pupal parasitoid of *Bruchus* spp., *Callosobruchus* spp., *Bruchidius atrolineatus*, and *Acanthoscelides obtectus* in legume seeds.

Although, biological control has a limited scope in stored grain management, it is becoming an important part of an IPM strategy. The main drawbacks of this method are it is very expensive and maintenance of culture is a must for insect pest control.

5.8 Botanicals

Keeping in view the discouraging aspects of synthetic pesticides such as toxicity, non-biodegradability, costlier, residual effects, and many other harmful effects on plants, humans, and other animals urged experts to look for an alternative powerful, economically viable, and eco-friendly approach. One such suitable method is the use of plant volatile organic compounds that possess insecticidal properties. Some plants are bestowed by nature with several bioactive organic chemicals or phytochemicals, having a defensive role against insect pests. These organic bioactive compounds provide an odor for the repellence of insects and are volatile in nature, hence commonly referred to as plant volatile organic compounds (PVOC). Plant volatiles is the most viable options for effective control measures against various pests, having no or fewer threats to the environment [37, 38]. Secondary metabolites of plants such as terpenoids, phenols, and alkaloids [39], act as attractants or repellents, influences the growth and development, ecdysis, fertility, behavior, mating, adult emergence, and plays an important role in crop protection [40]. Especially developing countries are using botanicals to reduce the infestation level [41]. Phytochemicals can be used in the form of aqueous or solvent extracts, powders, slurries, volatiles, and oils or shredded segments [38, 42]. Hence, botanicals hold promise as an alternative to synthetic insecticides to lessen the negative impact of the pesticide on the environment. Botanicals, as well as their active ingredients and the target pest upon which these are used, are listed in **Table 1** as enlisted by Singh et al. [43].

Botanicals or phytochemicals have a different mode of action on insect pests and consist of aldehydes, ketones, alcohols, alkanes, and terpenoids. Their effect on pests in several manners are described below:

Growth and development regulators: Phytochemicals are known to change the physiology and behavior of insects by affecting the growth, development, and metamorphosis of insects. Reduction in weight of larvae, pupa, and adult, prolonged larval and pupal periods are also some irreversible changes caused by botanical extracts [44]. Growth and development inhibition of C. maculatus is happened on applying the essential oil extracted from the *Cymbopogon schoenanthus* of the Poaceae family [45]. Botanicals have such a power of action that they can also inhibit the development of eggs and other immature stages residing inside the grain kernels. Aqueous extract of *Xanthium strumarium* leaf was reported to show toxicity, repellency, inhibition of fecundity and adult emergence of the pests, and grains as well as cereals protection against C. chinensis [46].

Hormone regulator: Plant volatiles has juvenile effects as well, i.e., they are playing an active role in the hormonal regulation of insect pests. Extracts of water hyacinth contain a juvenile hormone analog that changes the reproductive behavior and causes abnormal molting and metamorphosis of stored grain pests [47]. *Solasodine* could inhibit molting and induce several morphogenic abnormalities in the larvae of *T. confusum* at the concentration of 1 μg/μl.

Plant species	Family	Active ingredient	Target pest
Acorus calamus	Acoraceae	β-Asarone	*Sitophilus zeamais*
Aloysia citriodora	Verbenaceae	Citronellal and sabinene	*T. castaneum, T. confusum*
A. polystachya	Verbenaceae	Carvone and limonene	*T. castaneum, T. confusum*
Artemisia annua	Asteraceae	1, 8-cineole	*T. castaneum*
Baccharis salicifolia	Asteraceae	3-Carene	*T. castaneum, S. zeamais*
B. salicifolia	Asteraceae	β-Pinene	*T. castaneum, S. zeamais.*
Brugmansia suaveolens	Solanaceae	β-Pinene	*Zabrotes subfasciatus*
Carum carvi	Apiaceae	Carvone, Limonene, (E)-Anethole	*Rhyzopertha dominica, S. oryzae, S. zeamais*
Chamaecyparis obtusa	Cupressaceae	Bornyl acetate	*S. oryzae, C. chinensis*
Chenopodium ambrosioides	Amaranthaceae	Hexadecane	*T. castaneum, S. granarius*
Cinnamomum aromaticum	Lauraceae	Cinnamaldehyde	*T. castaneum, S. zeamais*
Citrus	Rutaceae	Limonene Eugenol	*T. castaneum, S. oryzae*
Colocasia esculenta	Araceae	2, 3-Dimethylmaleic anhydride	*S. oryzae, T. castaneum, C. chinensis*
Convolvulus arvensis	Convolvulaceae	Hexadecanoic acid	*R. dominica, S. oryzae*
Conyza dioscordis	Asteraceae	Dicotlyhexanedioate	*T. castaneum, S. granarius*
Coriander sativum	Apiaceae	Linalool	*S. oryzae, R. dominica and C. pusillus*
Cupressus lusitanica	Cupressaceae	Umbellulone and α-pinene	*T. castaneum, A. obtectus, S. cerealella and S. zeamais*
Duguetia lanceolata	Annonaceae	2,4,5-trimethoxystyrene	*Z. subfasciatus*
Eucalyptus spp.	Myrtaceae	α-Terpinene; 1, 8-Cineole; α-pinene	*S. oryzae*
Eucalyptus saligna	Myrtaceae	p-Cymene	*T. castaneum, S. oryzae*
Evodia ruticarpa	Rutaceae	Triterpenes	*T. castaneum, S. zeamais*
Feoniculum vulgare	Apiaceae	Phenylpropenes (E)-anethole Estragole (þ)-Fenchone	*S. oryzae, Lasioderma serricorne*
Juniperus foestidissima	Cupressaceae	Citronellol	*Trogoderma granarium*
Lantana camara	Verbanaceae	Coumaran	*S. oryzae, T. castaneum, R. dominica*

Plant species	Family	Active ingredient	Target pest
Melaleuca cajuputi	Myrtaceae	Terpine-4-ol Terpiniolene γ-Terpinene	*T. castaneum, S. oryzae, E. kuehniella, R. dominica*
Mentha citrate	Lamiaceae	Carvone, menthol, linalool, linalyl acetate	*T. castaneum, C. maculatus*
Nardostachys jatamansi	Caprifoliaceae	Aristolone	*T. castaneum, S. oryzae*
Ocimum canum	Lamiaceae	Linalool	*T. castaneum, S. granarius*
Ocimum kilimandscharium	Lamiaceae	Camphor	*S. oryzae*
Pimenta racemose	Myrtaceae	Linalool	*S. zeamais*
Rosmarinus officinalis	Lamiaceae	Camphor	*S. oryzae*
Spent hops	Lamiaceae	Xanthohumol	*S. granarius L., T. confusum and T. granarium*
Tagetes filifolia	Asteraceae	(E)-anethole and estragole	*T. castaneum*
Thespesia populnea	Malvaceae	Phenol	*C. maculatus*
Zingiber officinale	Zingiberaceae	1, 8-cineole	*T. castaneum, S. zeamais*
Z. officinale	Zingiberaceae	β-Zingiberene	*T. castaneum*

Source: Singh et al. [43].

Table 1.
Plant volatile organic compounds used against stored grain pests.

Oviposition deterrent: Chemicals that prevent or simply avoid insects from the process of oviposition is referred to as oviposition deterrent. Oviposition deterrents help to reduce the infestation level and offer the first line of defense against insect pests. An illusion is created by the plant volatiles to the gravid female pests, as these are involved in partially or completely preventing oviposition as well as the emergence of larvae from the laid eggs on stored grains [48, 49]. Garlic oil [50], 1,8 Cineole from Lamiaceae family [51], essential oils of *Eucalyptus citriodora*, *E. globulus*, *E. stageriana* [52], *Trachyspermum Ammi*, *Antheum graveolens*, *Nigella sativa*, are the oviposition deterrents, thereby reducing the viability of eggs and emergence of *Zabrotes subfaciatus*, *T. castaneum*, and *C. maculatus*. Finely powdered and dried leaves of *Ocimum* can completely suppress the oviposition of *Zabrotes subfascial* at 2%W/W, with an EC50, of 0.45% W/W [53]. The powdered form of *Chenopodium ambrosioides*, *Tagetesminuta*, *A. indica*, and *C. lusitanica*, applied at the rate of 1.5 kg/100 kg of *Phaseolus vulgaris*, was found to be the most effective in the mortality of Z. subfascial and A. objects [54]. Some phytochemicals obtained from *Laurus nobilis* and *Rosmarinus officinale* are causing egg mortality [55].

Repellent activity: Chemicals that protect stored grains, plants, or other products from insect damage by making the grains unattractive, offensive, or unpalatable to pests are commonly referred to as repellents. Repellents are especially more functional against various types of beetles, causing them to flee from the treated stored products. Compounds such as germacrol, pulegol, and α-terpineol isolated from *Baccharis salicifolia* and ar-turmerone isolated from *Curcuma longa* are potent repellents against *T. castaneum* and *S. zeamais* [56, 57]. Infestation by *T. castaneum* can be effectively controlled by different solvent extracts, acting as repellents, obtained from *Sphaeranthus indicus*, *Tephrosia purpurea*, *Prosopis juliflora*, *Cymbopogon flexuous, Cymbopogon winterianus*, and *C. martini*. Ethanolic extract of Acorus calamus is an active constituent Z-asarone, which acts as a strong repellent against *S. zeamais* [58]. Repellent used against *C. chinensis* is an essential oil obtained from *Callistemon lanceolatus* [59].

Antifeedant: Chemical substance that disrupts the feeding behavior of insect pests by making the treated stored grains unpalatable are referred to as antifeedant. The presence of certain chemicals in plants acts as a defensive mechanism to them. Antifeedants are eco-friendly, without ever disturbing the ecological balance, and do not kill the target but only prevent them from infestation. The deleterious effect of azadirachtin and neem seed extracts of *A. indica*, in antifeedant against various pests, is highly appreciable. Some essential oils acting as antifeedants are obtained from *Gaultheria procumbens*, against *S. oryzae* and *R. dominica* [60]. Some flavonoid compounds acting as antifeedants are Isoglabratephrin, −glabratephrin, tephroapollin-F, and lanceolatin-A, isolated from *Tephrosia apollinea*, against *T. castaneum*, *S. oryzae*, and *R. dominica*.

Ovicidal effects: Substances having the potential to kill eggs are considered to have ovicidal effects. This ovicidal property is also present in certain plants and is of great importance in the management of insect pests [61]. Plant volatiles sprayed on the stored grains could tremendously reduce the number of adult emergences because of toxicity or due to change in surface tension within the eggs [62]. Flavonoids isolated from *Calotropis Procera* provide 100% progeny suppression to the eggs of *C. chinensis* at 10 mg/ml concentration. An essential oil obtained from Anethum Sowa also shows ovicidal effects on eggs of *C. maculatus* [63]. From *Mentha ravens*, *Cinnamomum zeylanicum*, *Elettaria cardamomum*, *Syzygium aromaticum*, and *A. indica*, essential oils are extracted which has also ovicidal activity on the eggs of *T. castaneum* [64].

Chemosterilents: Substances that deprive insects of their ability to reproduce are known as Chemosterilents. Chemosterilents produce irreversible sterility without affecting the behavior of pests. These chemicals affect almost all stages of insect pests where eggs may not be oviposited, eggs not hatching, no pupation of larvae, and no adult emergence from these pupae [59]. Compounds possessing chemosterilent properties are asarone and 1,3,7-trimethylxanthine used against *C. chinensis* [65].

Behavioral disturbance: Behavioral changes are also induced by the plant volatiles, which can either stimulate or reduce insect mobility, and other physiological changes [66]. Some essential oils are known to inhibit acetylcholinesterase enzymes on insects' nervous system and also GABAergic is disrupted [67]. Essential oils of clove and Cinnamomum used on *S. zeamais* effects their locomotory and respiratory processes [68].

5.9 Pheromonal approach

Pheromones are ectohormones released by either male or female partners to change each other's behavior. Pheromones are now commercially available for around 20 species of stored grain pests [69]. Pheromones for *P. interpunctella*, *Lasioderma serricorne*, *T. castaneum, T. confusum, Trogoderma* variabile, are used frequently. These pheromones are placed inside suitable traps for their smooth release and maximum attraction as well as trapping processes. Proper installation of pheromone baited sticky traps within a building, granaries, and other flat landing sites plays an important role in the efficacy of pheromones used [70]. Sticky traps are placed on the sides of containers or the flat surface to capture crawling insects especially beetles, that eventually become stuck to the trapping surface. A trap with horizontal layers of corrugated cardboard was developed by [71] for beetles that walked through the tunnels of corrugations to reach a cup of oil into which they fell and suffocated.

6. Health and environmental hazards of pest control

Chemical insecticides are still considered as entomological weapons for the foreseeable future because of their wide host range, quick knockdown effects, easily availability to consumers. Their use in stored grain insect pests is still restricted as they pose threat to the health hazards and other environmental issues. Most of the chemical insecticides are carcinogenic and other health disorders [72]. The repeated application of insecticides leads to insecticide residues, secondary pest outbreaks. Recently the application of green synthesized nanoparticles is quite good and demonstrated satisfactory control against pulse beetle [73]. To overcome the issues of health and environmental hazards posed by chemical insecticides, workers are widely used other methods for their management. The satisfactory control has been observed when the product is not bulk and is being stored by physical and other methods which have been already discussed briefly in the chapter. Though, lot of botanicals have been applied to control a vast number of stored grain pests but satisfactory results are still wanted in large godowns especially in under-developed and developing countries. The villagers in these countries are still using the conventional methods and the damage levels are alarming and threatening. They even threaten the harvest which has been already harvested from different crops grown in the field and even protected conditions.

7. Conclusion

About 70% of stored grains are stored in villages in traditional methods. This creates an attractive atmosphere for the invasive pests to flourish. Especially developing countries have suffered a lot due to insect infestation. Integrated pest management is the best way to minimize the infestation status. Food supply to all human population, inhabiting any region of the world, seems less possible due to the alarming infestation rate of stored grains. IPM approach has many merits as it is the only method with which the quality as well as the number of stored products like grains, cereals, etc. are maintained to increase their economic value, as well as to provide nutritious food to even starved people.

Author details

Rayees Ahmad[1], Shafiya Hassan[1], Showkat Ahmad[2], Syed Nighat[1], Yendrambamb K. Devi[3], Kounser Javeed[2], Salma Usmani[5], Mohammad Javed Ansari[6], Sait Erturk[4], Mustafa Alkan[4] and Barkat Hussain[2]*

1 Entomology Research Laboratory, Department of Zoology, University of Kashmir, Srinagar, J&K, India

2 Division of Entomology, Department of Horticulture, SKUAST-Kashmir, India

3 Lovely Professional University, India

4 Plant Protection Central Research Instiute, Ankara, Türkiye

5 Department of Biochemistry, D.K.M. College for Women, Vellore, India

6 Department of Botany, Hindu College Moradabad, India

*Address all correspondence to: bhatbari@rediffmail.com

References

[1] Pruthi HS, Singh M. Pests of stored grain and their control. Special number. Indian Journal of Agricultural Science. 1950;**18**:1-52

[2] Khare BP. Pests of Stored Grain and their Management. New Delhi: Kalyani Publishers; 1994. pp. 304

[3] Pimentel D. World resources and food losses to pests. In: Gorham JR, editor. Ecology and Management of Food Industry Pests. Arlington, Virginia: Association of Official Analytical Chemists; 1991. pp. 5-11

[4] Morgan N, Aldred D. Post-harvest control strategies: Minimizing mycotoxins in the food chain. International Journal of Food Microbiology. 2007;**119**:131-139

[5] Jacobson M. Plants, insects and man their interrelationship. Economic Botany. 1982;**36**:346-354

[6] Atwal AS, Dhaliwal GS. Agricultural Pests of South Asia and their Management. New Delhi, India: Kalyani Publishers; 2008

[7] Shiferaw B, Prasanna BM, Hellin J, Bänziger M. Crops that feed the world past successes and future challenges to the role played by maize in global food security. Food Security. 2011;**3**(3): 307-327

[8] Bankole SA, Mabekoje OO. Occurrence of aflatoxins and fumonisins in preharvest maize from southwestern Nigeria. Food Additives and Contaminants. 2004;**21**(3):251-255

[9] Mihale MJ, Deng AL, Selemani HO, Kamatenesi MM, Kidukuli AW, Ogendo JO. Use of indigenous knowledge in the management of field and storage pests around Lake Victoria basin in Tanzania. African Journal of Environmental Science and Technology. 2009;**3**:49-71

[10] Emana G, Tsedeke A. Management of maize stem borer using sowing date at Arsi-Negele. Pest Management Journal of Ethiopia. 1999;**3**:47-52

[11] Tadesse A, Ayalew A, Getu E, Tefera T. Review of Research on Post-Harvest Pests. Abraham; 2008

[12] Baidoo PK, Mochiah MB, Owusu-Akyaw M. Levels of infestation on three different portions of the maize cob by the weevil Sitophilus zeamais (Motschulsky). Journal of Science and Technology (Ghana). 2010;**30**(3):6596

[13] Mason LJ, McDonough M. Biology, behavior, and ecology of stored grain and legume insects. Stored Product Protection. 2012;**1**:7

[14] Tefera T, Mugo S, Likhayo P. Effect of insect population density and storage time on grain damage and weight loss in maize due to the maize weevil, Sitophiluszeamais and the larger grain borer, Prostephanus truncates. African Journal of Agricultural Research. 2011;**6**:2249-2254

[15] Mebeasilassie A. Studies on the pest status of bean bruchid and managements of major bruchid species in central rift valleys of Ethiopia [doctoral dissertation, M.Sc. thesis]. Ethiopia: School of Graduate Studies, Addis Ababa University; 2004

[16] Adugna H, Dangnew G, Biniam Z, Biniam A. On farm storages studies in Eritrea. African Journal of Biotechnology. 2006;**5**(17):1537-1544

[17] Damte T, Dawd M. Chickpea, lentil and grass pea insect pest research in Ethiopia: A review. In: Food and Forage Legumes of Ethiopia: Progress and Prospects. Proceedings of a Workshop on Food and Forage Legumes. 2006. pp. 22-26

[18] Akob CA, Ewete FK. Laboratory evaluation of bioactivity of ethanolic extracts of plants used for protection of stored maize against Sitophilus zeamais Motschulsky in Cameroon. African Journal of Entomology. 2009;**17**(1): 90-94

[19] Allotey J, Morris JG. Biology of Cathartus quadricollis Guérin-Méneville Coleoptera: Silvanidae on some selected food media. Insect Science and Its Application. 1993;**14**:61-68

[20] Follett PA, Kawabata AM, Nelson R, Asmus G, Burt J, Goschke K, et al. Predation by flat bark beetles (Coleoptera: Silvanidae and Laemophloeidae) on coffee berry borer (Coleoptera: Curculionidae) in Hawaii coffee. Biological Control. 2016; **101**:152-158

[21] Via S. Cannibalism facilitates the use of a novel environment in the flour beetle, *Tribolium castaneum*. Heredity. 1999;**82**:267-275

[22] Rajashekar Y, Bakthavatsalam N, Shivanandappa T. Botanicals as grain protectants. Psyche: A Journal of Entomology. 2012:1-12

[23] Mobolade AJ, Bunindro N, Sahoo D, Rajashekar Y. Traditional methods of food. 2019

[24] Kesavan PC, Swaminathan MS. Strategies and models for agricultural sustainability in developing Asian countries. Philosophical Transactions of the Royal Society B: Biological Sciences. 2008;**363**:877-891

[25] Suleiman M, Rugumamu CP. Management of insect pests of stored sorghum using botanicals in Nigerian traditional stores. Journal of Stored Products and Postharvest Research. 2017;**8**:93-102

[26] Dales MJ. A Review of Plant Materials Used for Controlling Insect Pests of Stored Products. NRI; 1996. p. 1996

[27] Casida J. Pyrethrum: The Natural Insecticide. Elsevier; 2012

[28] Dayan FE, Cantrell CL, Duke SO. Natural products in crop protection. Bioorganic & Medicinal Chemistry. 2009

[29] Smith RL, Adams TB, Doull J, Feron VJ, Goodman JI, Marnett LJ, et al. Safety assessment of allylalkoxybenzene derivatives used as flavouring substances-methyl eugenol and estragole. Food and Chemical Toxicology. 2002;**40**(7):851-870

[30] Erturk S. Combined and individual effects of diatomaceous earth and methyl eugenol against stored products insect pests. Turkish Journal of Entomology. 2021;**45**(2):173-184

[31] Isman MB. Botanical insecticides, deterrents, and repellents in modern agriculture and an increasingly regulated world. Annual Review of Entomology. 2006;**51**:45-66

[32] Fields PG, White ND. Alternatives to methyl bromide treatments for stored product and quarantine insects. Annual Review of Entomology. 2002;**47**:331-359

[33] Norman CS. Potential impacts of imposing methyl bromide phaseout on US strawberry growers: A case study of a nomination for a critical use exemption under the Montreal Protocol. Journal of Environmental Management. 2005;**75**:167-176

[34] Horn F, Horn P, Sullivan J. Current practice in fresh fruit fumigation with phosphine in Chile. In: Proceedings of 2005 Annual Research Conference on Methyl Bromide Alternatives and Emissions Reductions. 2005. p. 31

[35] Flinn PW, Kramer KJ, Throne JE, Morgan TD. Protection of stored maize

from insect pests using a two-component biological control method consisting of a hymenopteran parasitoid, *Theocolax elegans*, and transgenic avidin maize powder. Journal of Stored Products Research. 2006;**42**:218-225

[36] Grieshop MJ, Flinn PW, Nechols JR, Schoeller M. Foraging success of three species of *Trichogramma* (Hymenoptera: Trichogrammatidae) in a simulated retail environment. Journal of Economic Entomology. 2007;**100**:591-598

[37] Isman MB. Botanical insecticides: For richer, for poorer. Pest Management Science. 2008;**64**:8-11

[38] Rajashekar Y, Tonsing N, Shantibala T, Manjunath JR. 2, 3-Dimethylmaleic anhydride (3, 4-Dimethyl-2, 5-furandione): A plant derived insecticidal molecule from Colocasia esculenta var. esculenta (L.) Schott. Scientific Reports. 2016;**6**:20546

[39] Rajashekar Y, Rao LJ, Shivanandappa T. Decaleside: A new class of natural insecticide targeting tarsal gustatory sites. Naturwissenschaften. 2012;**99**:843-852

[40] Bennett RN, Wallsgrove RM. Secondary metabolites in plant defence mechanisms. The New Phytologist. 1994;**127**(4):617-633

[41] Rajashekar Y, Gunasekaran N, Shivanandappa T. Insecticidal activity of the root extract of *Decalepis hamiltonii* against stored-product insect pests and its application in grain protection. Journal of Food Science and Technology. 2010;**47**:310-314

[42] Rajashekar Y, Bakthavatsalam N, Shivanandappa T. Botanicals as Grain Protectants. Psyche: A Journal of Entomology. 2012. Article ID 646740

[43] Singh KD, Mobolade AJ, Bharali R, Sahoo D, Rajashekar Y. Main plant volatiles as stored grain pest management approach: A review. Journal of Agriculture and Food Research. 2021;**4**:1-12

[44] Rosenthal GA, Dahlman DL, Crooks PA, Phuket SN, Trifonov LS. Insecticidal properties of some derivatives of L-canavanine. Journal of Agricultural and Food Chemistry. 1995;**43**:2728-2734

[45] Ketoh GK, Koumaglo HK, Glitho IA. Inhibition of *Callosobruchus maculatus* (F.) (Coleoptera: Bruchidae) development with essential oil extracted from *Cymbopogon schoenanthus* L. Spreng. (Poaceae), and the wasp *Dinarmus basalis* (rondani) (Hymenoptera: Pteromalidae). Journal of Stored Products Research. 2005;**41**:363-371

[46] Roy B, Amin MR, Jalal S, Kwon YJ, Suh SJ. Evaluation of common cocklebur *Xanthium strumarium* leaf extract as post-harvest grain protectant of black gram against pulse beetle *Callosobruchus chinensis* (Coleoptera: Bruchidae) and isolation of crude compound. Entomological Research. 2014; **44**:254-261

[47] Rani PU, Jamil K. Effect of water hyacinth leaf extract on mortality, growth and metamorphosis of certain pests of stored products. International Journal of Tropical Insect Science. 1989;**10**:327-332

[48] Hora KH, Roessingh P. Oviposition in Yponomeuta cagnagellus: The importance of contact cues for host plant acceptance. Physiological Entomology. 1999;**24**:109-120

[49] Sousa RMO, Rosa JS, Oliveira L, Cunha A, Fernandes-Ferreira M. Activities of Apiaceae essential oils and volatile compounds on hatchability, development, reproduction and nutrition of *Pseudaletia unipuncta* (Lepidoptera: Noctuidae). Industrial Crops and Products. 2015;**63**:226-237

[50] Yang FL, Zhu F, Lei CL. Insecticidal activities of garlic substances against adults of grain moth, *Sitotroga cerealella* (Lepidoptera: Gelechiidae). Insect Science. 2012;**19**:205-212

[51] Koschier EH, Sedy KA. Effects of plant volatiles on the feeding and oviposition of Thrips tabaci. Thrips and Tospoviruses. 2001:185-187

[52] Tapondjou LA, Adler CLAC, Bouda H, Fontem DA. Bioactivities of cymol and essential oils of *Cupressus sempervirens* and *Eucalyptus saligna* against *Sitophilus zeamais* Motschulsky and *Tribolium confusum* du Val. Journal of Stored Products Research. 2005; **41**:91-102

[53] Weaver DK, Dunkel FV, Potter RC, Ntezurubanza L. Contact and fumigant efficacy of powdered and intact *Ocimum canum* Sims (Lamiales: Lamiaceae) against *Zabrotes subfasciatus* (boheman) adults (Coleoptera: Bruchidae). Journal of Stored Products Research. 1994; **30**(3):243-252

[54] Paul UV, Lossini JS, Edwards PJ, Hilbeck A. Effectiveness of products from four locally grown plants for the management of *Acanthoscelides obtectus* (Say) and *Zabrotes subfasciatus* (Boheman) (both Coleoptera: Bruchidae) in stored beans under laboratory and farm conditions in Northern Tanzania. Journal of Stored Products Research. 2009;**45**(2): 97-107

[55] Isikber AA, Alma MH, Kanat M, Karci A. Fumigant toxicity of essential oils from *Laurus nobilis* and *Rosmarinus officinalis* against all life stages of *Tribolium confusum*. Phytoparasitica. 2006;**34**:167

[56] García M, Donadel OJ, Ardanaz CE, Tonn CE, Sosa ME. Toxic and repellent effects of *Baccharis salicifolia* essential oil on *Tribolium castaneum*. Pest Management Science. 2005;**61**:612-618

[57] de Souza Tavares W, de Sousa Freitas S, Grazziotti GH, Parente LML, Liao LM, Zanuncio JC. Ar-turmerone from *Curcuma longa* (Zingiberaceae) rhizomes and effects on *Sitophilus zeamais* (Coleoptera: Curculionidae) and *Spodoptera frugiperda* (Lepidoptera: Noctuidae). Industrial Crops and Products. 2013;**46**:158-164

[58] Yao Y, Cai W, Yang C, Xue D, Huang Y. Isolation and characterization of insecticidal activity of (Z)-asarone from *Acorus calamus* L. Insect Science. 2008;**15**(3):229-236

[59] Shukla R, Singh P, Prakash B, Kumar A, Mishra PK, Dubey NK. Efficacy of essential oils of *Lippia alba* (Mill.) NE Brown and *Callistemon lanceolatus* (Sm.) Sweet and their major constituents on mortality, oviposition and feeding behaviour of pulse beetle, *Callosobruchus chinensis* L. Journal of the Science of Food and Agriculture. 2011;**91**(12):2277-2283

[60] Kiran S, Prakash B. Assessment of toxicity, antifeedant activity, and biochemical responses in stored-grain insects exposed to lethal and sublethal doses of *Gaultheria procumbens* L. essential oil. Journal of Agricultural and Food Chemistry. 2015;**63**:10518-10524

[61] Hong TK, Perumalsamy H, Jang KH, Na ES, Ahn YJ. Ovicidal and larvicidal activity and possible mode of action of phenylpropanoids and ketone identified in *Syzygium aromaticum* bud against *Bradysia procera*. Pesticide Biochemistry and Physiology. 2018;**145**:29-38

[62] Singh SR, Luse RA, Leuschner KE, Nangju D. Groundnut oil treatment for the control of *Callosobruchus maculatus* (F.) during cowpea storage. Journal of Stored Products Research. 1978;**14**(2-3): 77-80

[63] Tripathi AK, Prajapati V, Aggarwal KK, Kumar S. Insecticidal and ovicidal activity of the essential oil of

Anethum sowa Kurz against *Callosobruchus maculatus* F. (Coleoptera: Bruchidae). International Journal of Tropical Insect Science. 2001;**21**(1): 61-66

[64] Aggarwal KK, Tripathi AK, Ahmad A, Prajapati V, Verma N, Kumar S. Toxicity of L-menthol and its derivatives against four storage insects. International Journal of Tropical Insect Science. 2001;**21**(3):229-235

[65] Rizvi SJH, Pandey SK, Mukerji D, Mathur SN. 1, 3, 7-Trimethylxanthine, a new chemosterilant for stored grain pest, *Callosobruchus chinensis* (L.). Journal of Applied Entomology. 1980;**90**:378-381

[66] de Araujo AMN, Faroni LRDA, de Oliveira JV, Navarro DMDAFB, MO, de França SM. Lethal and sublethal responses of *Sitophilus zeamais* populations to essential oils. Journal of Pest Science. 2017;**90**(2):589-600

[67] Bloomquist JR, Boina DR, Chow E, Carlier PR, Reina M, Gonzalez-Coloma A. Mode of action of the plant-derived silphinenes on insect and mammalian GABAA receptor/ chloride channel complex. Pesticide Biochemistry and Physiology. 2008; **91**(1):17-23

[68] Haddi K, Oliveira EE, Faroni LR, Guedes DC, Miranda NN. Sublethal exposure to clove and cinnamon essential oils induces hormetic-like responses and disturbs behavioral and respiratory responses in *Sitophilus zeamais* (Coleoptera: Curculionidae). Journal of Economic Entomology. 2015;**108**(6):2815-2822

[69] Phillips TW, Throne JE. Biorational approaches to managing stored-product insects. Annual Review of Entomology. 2010:1-12

[70] Nansen C, Phillips TW, Sanders S. The effect of height and adjacent surfaces on captures of the Indianmeal moth, *Plodia interpunctella* (Lepidoptera: Pyralidae), in pheromone-baited traps. Journal of Economic Entomology. 2004; **97**:1284-1290

[71] Barak AV, Burkholder WE. A versatile and effective trap for detecting and monitoring stored product Coleoptera. Agriculture, Ecosystems and Environment. 1985;**12**:207-218

[72] Ahmad R, Hussain B, Ahmad T. Fresh and dry fruit production in Himalayan Kashmir, Sub-Himalayan Jammu and Trans-Himalayan Ladakh, India. Heliyon. 2021;7(1):e05835

[73] Rahman MA, Parvin A, Khan MSH, War AR, Lingaraju K, Prasad R, et al. Efficacy of the green synthesized nickel-oxide nanoparticles against pulse beetle, *Callosobruchus maculatus* (F.) in black gram (*Vigna mungo* L.). International Journal of Pest Management. 2020:1-9

Section 3

Postharvest Disease Management of Fruits and Vegetables

Chapter 4

Robotic Heat Treatments for Mango and Prickly Pear Increase Shelf Life and Reduce Pathogen Infection

Federico Félix Hahn Schlam

Abstract

Mexico is the main exporter of mango fruits and prickly pears, so new postharvest techniques to increase shelf life are studied. Thermal treatments on both fruits can affect their cuticle so it was reviewed. When mango latex remains within the fruits, it avoids sap burn and decreases anthracnose and stem end rot infestation, so two systems were developed to minimize latex de-sapping. A gripper cuts stems 0.5 cm long and cauterizes them with a hot knife implement. A heating gun applied paraffin wax to mangoes without the stem end and protected them better against anthracnose lesions. Physicochemical analysis of several mango varieties was carried out after harvesting, at market place and after pedicel cutting and cauterizing. Keitt mangoes showed the lower quantity of total soluble solids (TSSs) and total acidity (TA). When the pedicel was cauterized, TSS dropped. Two grippers were developed to cryo-cauterize prickly pears as this system is more energy-efficient than hot cauterization. A six-finger gripper moved over a pneumatic actuator toward a dry ice chamber to optimize pear cryo-cauterization. Gripper's strong grasping damaged the fruits due to excessive compression. TSS and TA of cryo-cauterized fruit remained constant during the three months of fruit storage.

Keywords: mango fruits, anthracnose, grippers, prickly pear, paraffin wax, cryo-cauterization, total soluble solid concentration, stem end rot

1. Introduction

Every country develops studies for their main fruit chains to determine main losses and provide solutions for reducing them. When fruit shelf life cannot be increased, processing will avoid fruit spoilage. Food losses and waste are estimated globally in 1.3 billion tons annually. Commercialization loss was estimated in 9.5 tons/week in Salvador, Brazil, in highly perishable fruits such as banana, papaya, and tomato [1]. The annual loss of fruits during postharvest operation represents in Sri Lanka about 210,000 metric tons of fruit, which corresponds to 30–40% of the harvest, representing a loss of US$90 million [2]. Mexico is the leading producer of prickly pear plants with 230,000 hectares, being 67,000 for fruit production [3]. Mexico is also the world leader in exporting fresh mangoes in 2019 [4]. Postharvest losses of fresh mango fruits in Pakistan were reported to average 69% [5] but sometimes reach 100% under disease-favorable environments. In the 2014 season, an increase in mango stem end rot (SER) at Israel caused a 30–40% loss of the harvested fruit [6]. This disease occurs in mango, avocado, and citrus fruit [7].

The rind or exocarp includes the hard cases of nuts or the shell of watermelon. The peel forms the pericarp, meanwhile the pulp or edible portion of the fruit is the endocarp [8]. Fruit or vegetable peel or rind appears as its outer protective layer. Watermelon, a round fruit, has a firm outer rind that surrounds a white inner rind layer. The interior edible pulp of red or yellow color is the endocarp. The outer walls of the epidermal cells of all plant organs are coated with a cuticular membrane [9]. Physical properties and chemical composition of the fruit cuticle change markedly during its development [10]. During early fruit development, maximum cuticle deposition rates per unit area appear increasing cuticle thickness. Cuticle composition changes after depositions of wax, phenolic compounds, and polysaccharides [11].

Fleshy fruit cuticles and vegetative organs have similar compounds, but fruit cuticles are thicker [12, 13]. The hydrophobic nature of fruit cuticle makes it an effective barrier to reduce water loss. Cuticle permeance differs between mango fruits receiving sunshine and those growing under the canopy shade [14]. In addition, intracuticular waxes limit movement of surface water into the fruit and reduce transpiration. Cuticular wax load increases during fruit development leading to a thicker mango cuticle at maturity [15].

The fruit cuticle provides an important physical barrier against pathogens [16] avoiding fungal colonization on sweet oranges [17]. Industrial food wastes such as peels from juice production provide raw material for obtaining wax compounds [18]. The cuticle also provides protection against environmental conditions, where excessive solar radiation produces physiological disorders such as sunscald [19]. Cuticle strength and rigidity decrease as it becomes warmer [20]. The cuticle inner surface is fully hydrated, meanwhile the cuticle outer surface in contact with the atmosphere is less hydrated. Although waxes are present in both sides of the cuticle, water absorption takes place [10]. Cuticle swelling and softening alter its mechanical properties. Fruit cracking is triggered by cuticle breaking, linked to rainwater and high humidity [21, 22].

Handling fruits up to 15 days after harvest has a profound effect on its final quality because fruits are still alive and vulnerable to adverse conditions [23]. Throughout fruit ripening, softening results from the modification of polymers within the primary cell wall [24]. Cuticle and wax deposition increased during the first 15 days of postharvest shelf life in mango fruits of cultivars "Kent," "Tommy Atkins," and "Ataúlfo" [25]. Mango fruits with higher wax deposition in their cuticle were more resistant to fruit fly attack [26]; also fruits presented lower transpiration and deterioration. Pectin solubilization during fruit ripening is directly related with the ripe fruit texture [27]. Fruits showing a melting texture, such as avocado, kiwifruit, tomato, and peach, soften in a short time [28]. Fruits having a crispy texture during maturation, such as apple or watermelon, present low pectin solubilization [29]. The simplest postharvest procedure to increase fruits shelf life consists of storing them under controlled temperature and humidity conditions. However, rheological and mechanical properties of fruit cuticles are affected [20]. Peach firmness dropped after being stored at low temperatures. It was associated to a reduction of covalently bound pectins [30]. Apricot controlled-atmosphere treatments showed also pectin degradation [31].

Mango fruit pedicel (**Figure 1a**) presents an internal network of resin ducts, and the latex is kept under plant turgor pressure [32]. When the pedicel is broken or cut, a secretion of milky-viscous sap leaves the fruit [33]. This latex contains oily antifungal resorcinol [34]. The contact of the fruit surface with the sap exudate (**Figure 1c**) can

Figure 1.
Mango fruit (a) pedicel, (b) washing, and (c) showing latex in the peel.

lead to skin injury (sap burn) and develop under-skin browning [32]. This injury decreases mango quality after damaging seriously its skin, and if the fruit contacts the soil, it can be easily infected. These fruits are rejected at the entrance of fresh fruit packinghouses [35]. Lenticels also appear after sap exudation showing symptoms of early sap burn injury [36]. A delay in the appearance of stem end rot was noted by keeping a short pedicel at harvest [6, 34]. Mango fruits harvested with stems have more sap and less incidence of anthracnose [37].

Opuntia species present fleshy edible fruits (tunas) with thick rinds and relatively large round seeds. Prickly pear fruits are consumed in the local Mexican market and exported to the United States, Canada, Japan, and Europe [38, 39]. Edible prickly pear fruits and cladodes are used as food for livestock [40]. Fruit pulp and peel present a high quantity of carotenoids, betalain content, polyphenolic content, and ascorbic acid [41, 42]. Those pigments have revalorized prickly pear production for agro industries and pharmaceutical use [3, 43]. The fruit is perishable, and after being stored for nine days at room temperature, it starts rotting [44]. Ready-to-eat (RTE) fruit storage includes controlled atmosphere storage of minimally processed cactus pear fruits at 2°C reducing browning content [45]. Cactus pear peeled and stored within passive-modified atmosphere at low temperature limited fruit decay [46].

Heat transfer within fruits stored at a cold storage warehouse after harvesting has been studied before long-term shipping [47–49]. Harvested fruits are treated with different technologies to delay ripening, preventing physiological and pathological disorders [49]. Producers sometimes target distant markets, so they must harvest their tomatoes in a mature green state to allow longer ripening and senescence periods [50]. Excessive field heat increases fruit metabolic activity, so immediate cooling after harvest is recommended [51]. Low and high temperatures lead to the denaturation of enzymes, modifying fruit's respiration rate [52]. Stone fruits such as plums and mangoes have a seed inside and present different thermophysical parameters within the pulp [53, 54]. The contact surface between the seed and the pulp is the deepest point that can be reached in the fruit and becomes a thermal center. The finite element method can simulate heat transfer within food products that present irregular geometries [55].

Hot water immersion and hot air treatments at temperatures between 40 and 60°C from seconds to several hours control pathogens in apples, pears, citrus, and melons [56]. Postharvest quality of apples improved after being heated with air during one day at 40°C [57]. Heat treatment caused important changes in epicuticular wax altering microcrack structure. Mandarins were immersed in hydrothermal treatments, maintaining the fruit surface temperature at 50°C for 2.5 min [58]. Once the mandarin peel heats up, thermal energy transfers by conduction to subsequent layers toward the center. Heat transfer stops after reaching an equilibrium condition [59]. Thermally treated mandarins present higher TSS (total soluble solids), lower maturity index, and similar citric acid content.

Mango fruit must be treated to ensure that it is free of fruit flies, so that importing markets allow their acceptance [60]. Small mango fruits weighing less than 375 g require 65 min of immersion in hot water at 46.1°C [61]. A thermocouple was inserted at the surface of the endocarp and another in the center of the mango fruit to record temperature changes during hot water immersion. The temperature at the center of the fruit continued increasing for 10 min after removing the fruit from the hot water bath [61]. Although the hot water treatment reduces fruit firmness, it influences positively in oxidative processes, cell wall changes, and steady-state levels of protein [62].

2. Mango treatment

Thermal treatment application maintains mango fruit quality and produces higher economic returns. Cauterization is a very useful technique that can close any tissue after applying heat. After harvesting, all the wounds of the fruit that were cauterized and sealed hermetically avoiding transpiration and increasing shelf life.

2.1 Mango after farm harvest

Postharvest mango quality depends on proper harvesting and even better production practices. Mangoes are generally handpicked or retrieved with poles adapted with a cutting blade and a bag [63]. The blade end breaks the pedicel and latex covers the fruit peel (**Figure 1c**). Although de-sapping after harvest avoids peel sap burn, it reduces fruit protection against anthracnose and stem end rot. The main cause of mango sap burn is attributed to a deposit of volatile compounds such as terpinolene and car-3-ene through the lenticels [64]. Stem trim cutting results in latex stains deposited on the fruit surface. The sap stored in the fruit ducts under high pressure falls on the peel of mango fruit [65]. Delatexing can be done by inverting freshly de-stemmed fruits on plastic or steel mesh racks for 30 min. Another technique is to dip freshly de-stemmed fruits in 1% alum solution (one-half kg powdered alum per 50 L of water) for 1 min; fruits should dry before packing [65]. The contact of latex with mango skin induces lenticel discoloration, resulting in red spots caused by the synthesis of anthocyanins [66]; these spots can also be induced by chilling injury [64]. Resorcinols and gallotannins are inhibitory to major postharvest pathogens including anthracnose [67].

If a 1 cm long pedicel remains attached to the fruit after harvest, latex will not leave the fruit avoiding sap burn. More than 80% of sap flow was observed within the first minute of stalk removal [37]. Sap pH varies between 4.43 and 4.6, and the ratio of nonaqueous fluid (oil) to aqueous fluid is of 1:6.5 [37]. The best hour for harvesting mango fruits was just after midday [68]. Early morning harvesting causes a rapid flow of sap from the pedicel end. High solar radiation and vapor pressure deficit increased stem water flow within mature fruit during the morning and decreased after midday [69]. Pedicel cutting place does not affect sap output flow. If stem is cut at the abscission zone, delayed ripening of mango fruit results [68].

2.2 Mango diseases

Two of the main diseases of mango fruits are anthracnose and stem end rot. Anthracnose caused by the *Colletotrichum gloeosporioides* at the green stage cannot be perceived, and the infection is noted when the mango ripens. Anthracnose produces the enzymes polygalacturonase and pectolyase, which degrade the cell wall [70]. If mango fruit is healthy, the polyphenol oxidase (PPO) enzyme is found within chloroplasts and the phenolic compounds in vacuoles, both being separated, avoiding any reaction.

Stem end rot (SER) is a disease caused by *Lasiodiplodia theobromae*. At the beginning, it appears as a small dark-brown area in the peel around the base of the fruit stem end, progressing into soft decay at the stem end [6]. Ethylene, a phytohormone, controls most of the ripening events linked with climacteric fruits. Small amounts of ethylene maintain fruit resistance to pathogens [71].

2.3 Mango pedicel treatment

If latex is retained within the fruit at harvest, it reduces anthracnose and stem end rot (SER) development during ripening. Fruit ripening parameters are not affected by pedicel length, and substantially less number of diseases appear compared with fruits harvested without stems. Anthracnose lesions decrease when mango fruit is harvested with a long stem [33]. SER onset in fruits with short pedicel was later than in fruit without stems [6]. Latex aqueous phase having chitinase contributes to fruit resistance against SER [67]. Two systems were developed to minimize latex de-sapping:

1. Cut stems 0.5 cm long and cauterize them with a hot knife implement.

2. If harvest brings fruits without stems, fruits are washed, dried and a wax is applied at the stem end.

Automatic fruit harvesting follows different picking patterns including bending, lifting, twisting, and pulling [72]. Modern soft grippers employ soft and flexible materials for holding the fruits [73]. Mechanical cutting devices for fruits consist of knifes [74, 75], scissors [76], and hot wires [77]. Knives used to cut stems have to be continuously immersed in skimmed milk. This action avoids virus invasion and should take place before contacting each plant. Therefore, it is not practical for automated processes [74]. A scissor employed to cut tomato stalks was articulated by a finger phalanx, but could also be fixed to the gripper palm [76].

Nichrome wire electrodes were mounted at a thermal cutting end effector. A high voltage of 300 V cuts 1 mm sweet pepper stems in 2 s [77]. As the diameter doubled, the cutting period increased to 5 s after applying the same voltage between electrodes [77]. Thermal stem cut ceased fungal or bacterial infestation, increasing pepper shelf life over 15 days. Peppers harvested by normal scissors showed physical changes after the fifth day and perished after nine days. Mechanical cutting is suitable for cucumbers where peduncle direction is uniform [74]. Laser cutting of variable-diameter tomato peduncles (1.5–5 mm) was studied [78]. It became impossible to cut off a peduncle directly by focusing a laser beam on it, as the focusing spot is smaller than the peduncle size. After tomato peduncle drilling, laser cut a 5 mm diameter stem in 15.2 s [78].

A harvesting robot requires a transmission system to drive the end effector [79]. A robot gripper with four pneumatic fingers has been used with mango fruits. The gripper can handle various shapes and sizes and has been used to determine fruit firmness [79]. A gripper was also developed to handle mango fruits and estimate their ripeness. This robot integrated accelerometers and optical sensors and worked without contacting the fruit [80]. Two robots were used for tomato grafting, cutting 240 plants per hour. The graft is accomplished when both plants are placed in intimate contact between them, and a clip is pressed against them [81].

Mango fruits collected at the Mexican Pacific coast were green, firm, and starting to ripe. The developed gripper to hold the fruit presented integrated soft cushions (**Figure 3(a** and **b)**) to protect the fruit and move it for cutting the stem. Two linear knifes were used by the trimmer equipment. One knife was fixed, meanwhile the other was ejected by a 24 VDC (direct current voltage) linear actuator. Preliminary tests show successful results in stem cutting with only one movement. The mango enters the transporting system, but not all the fruits have attached pedicels. Those having the pedicel were cut by a warm knife having a temperature of 35°C. An image of the mango peduncle or abscission zone was obtained with a X800 digital microscope. The effect of anthracnose infestation was analyzed after fruit matured.

Wax was applied to mango fruits without the stem end. Paraffin was warmed up in the interior of a conventional gun (**Figure 3a**) and applied to the mango abscission orifice to avoid fungal or bacteria infestation. The manual gun uses paraffin sticks that melt after being heated by an electric resistance. When the trigger is squeezed, liquid wax leaves the gun through an output nozzle. Better results are obtained after applying pressure with a conical stamp over the liquid wax placed at the fruit peduncle orifice (**Figures 1a** and **2b**). An industrial wax application gun pressurizes the hot fluid with a pneumatic system (**Figure 3b**). A camera at the top provides information of whether the fruit has a 1 cm long stem and would only apply wax when there is no pedicel.

2.4 Mango pedicel and abscission microscope images

Large latex channel openings were seen at and below the abscission zone close to the fruit. High volume of latex spurts out through these channels after detaching the pedicel from the fruit [82]. Latex canals are seen as large perforations in the fruit peel reaching the outer pulp [34, 82]. After cutting the Keitt mango pedicel 2 cm away from the abscission zone, it was cauterized at 35°C, showing latex channels (**Figure 4a**). Cauterization at 35°C does not heat mango peel tissue (**Figure 4b**). If the stalk was cauterized at 45°C, the cells surrounding the channels were burnt and reduced in size (**Figure 4c**). Latex channels are clearly observed within red

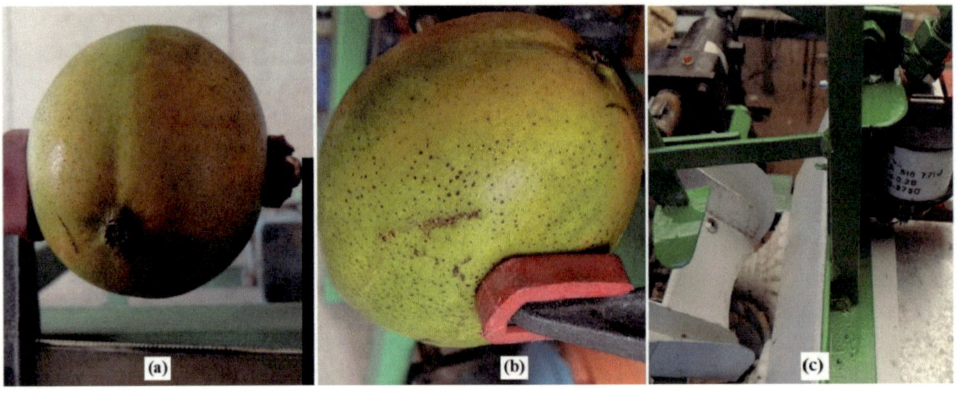

Figure 2.
Robotic gripper (a) with mango pedicel, (b) without mango pedicel and having wounds, (c) cauterizer knife machine.

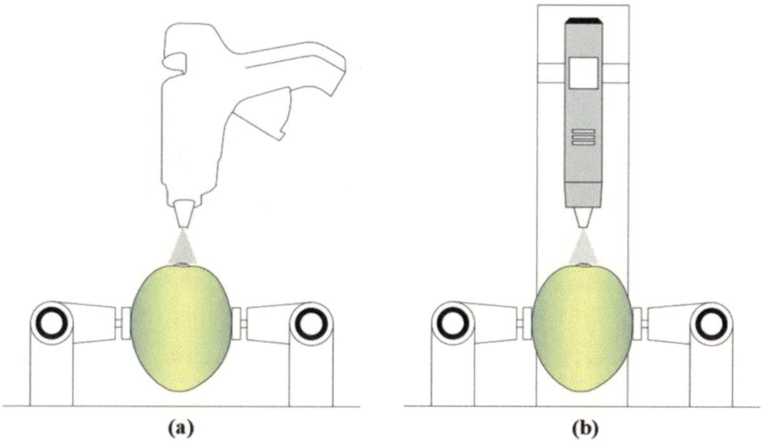

Figure 3.
Robotic arms handling a mango fruit for (a) manual, and (b) industrial wax application.

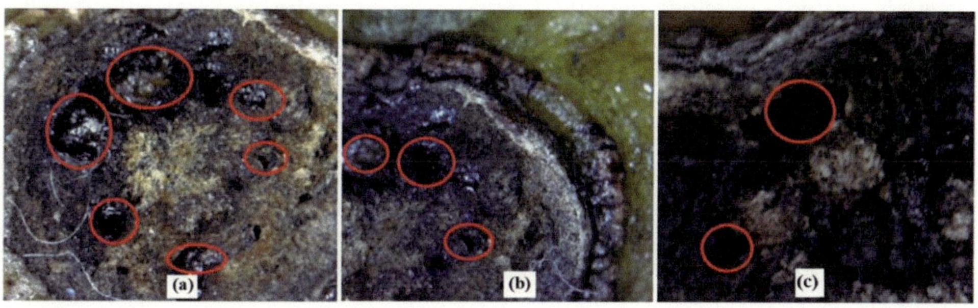

Figure 4.
Transverse section of Keitt mango fruit stem, showing the latex canals after cauterization at (a) 35°C on the abscission zone, (b) 35°C on the pedicel, and (c) 45°C on the pedicel.

circles in the green tissue just after removing the pedicel (**Figure 5a**). If the stem gets cauterized, latex channels are still present after cutting the pedicel with a razor blade, 0.5 cm toward the fruit abscission end (**Figure 5b**). If honey covers the green tissue, it will enclose the latex channels (**Figure 5c**).

3. Prickly pear treatments and measurements

Cactus pear (*Opuntia ficus-indica L.*) is an important fruit, but its consumption is limited by the presence of spines and glochids on its surface. Fresh-cut, ready-to-eat (RTE) cactus pears have higher preference than the whole fruits [83]. Actually, cactus pear at the green-yellow ripening stage is processed as a ready-to-eat fruit and stored for nine days in modified atmosphere packaging at 4°C [84]. Green yellow fruits present intermediate peel thickness and pulp softness, which is suited for peeling and for RTE fruits [85].

Cauterization prototypes were developed to increase prickly pear shelf life and decrease fruit diseases. A review on cauterization techniques including high-temperature contact and cryo-cauterization was presented [38]; both of these systems are patented [86, 87]. A cauterizer for harvested fruits applied 100 kPa of pressure at 200°C during 30 s [88]. Cactus pears subjected to a cauterization treatment were cut at the top-peduncle section, leaving a sealing area of 13 cm^2. The system is efficient in controlling postharvest diseases, but its excessive heat application

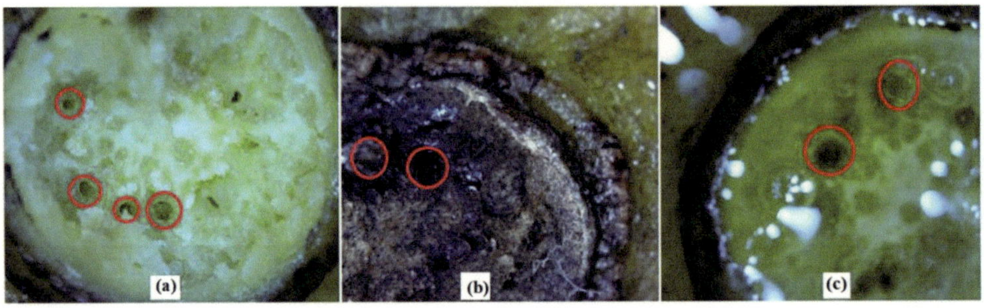

Figure 5.
Transverse section of Keitt mango showing the latex canals after (a) removing the pedicel, (b) cauterizing, and (c) removing the stem and adding honey.

results in expensive energy consumption [88]. Pulp temperature increased to 86°C after heating the fruit at 200°C for 45 s [88].

Prickly pear and their cladodes have natural polymers, and several eco-friendly materials are under development [89]. Cactus mucilage can be used as gelling, stabilizing, or encapsulating agent. The use of this bio-polymer material opens new opportunities in the food packaging. It is also used as a flocculating agent for heavy metals in water [90]. All these properties open new economic opportunities for cactus produce.

3.1 Prickly pear automatic cold cauterization

Several mechanisms have been developed for detaching the fruit from the cladode [91] and for fruit cold cauterization [92]. A harvesting arm with four degrees of freedom is used as hydraulic piston to collect prickly pears [91]. Cryo-cauterization results from pressing the fruit sliced area against a dry ice wall. The thermal shock maintained cactus pear over 120 days without further cooling [44]. Energy consumption of cryo-cauterization was minimum as no resistance was used; meanwhile the cauterizer working at 200°C employed 13 W per fruit [88]. The first automatic fruit cauterizer uses sensors and mechanisms to detect when the prickly pear is present within the metal container, rotate it 90° counterclockwise, displace it against the dry ice wall and deposit it again into the original band. The processing of 1000 fruits took a little more than 500 min [92]. Further development to simplify the system used a two-finger gripper that picks the fruit (**Figure 6a**). The most significant features to select a gripper include its opening range, its maximum applied force, its type of movement (angular, parallel or self-centered), and the grasp strategy (external or internal grasp). The robotic end effector uses two fingers to press the thick fruit peel without damaging it. The mechanism rotates the fruit by180° until it touches the ice pad (**Figure 6b**). However, dry ice melts in 5 h and has to be replaced in both systems. The last prototype has a gripper that grasps the fruit more efficiently with six fingers (**Figure 6c**). The gripper moves horizontally toward the dry ice chamber by sliding on pneumatic actuators. In the slider actuators, the gripper is mounted to the carriage. Precise slicing of the top-peduncle section is done by means of a circular blade. Once the fruit is sliced, it moves further to the left until it presses the dry ice chamber. With additional volume of dry ice within the chamber, it can last more than one day.

3.2 Prickly pear temperature measurements

Thermocouple sensors are being used for monitoring temperature within the fruit. Sensors were added below mesocarp and in the center of the fruit to study fruit changes during hot water treatments [58]. Three thermocouples of type J were

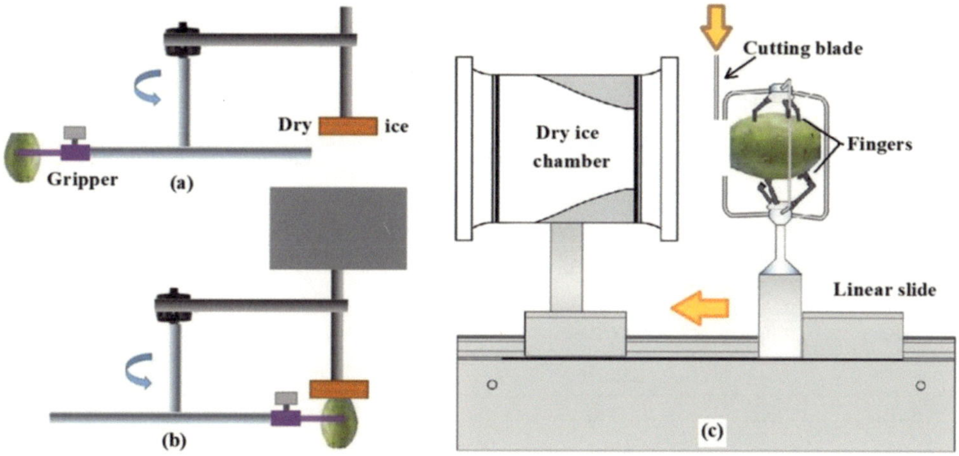

Figure 6.
Rotating gripper (a) picking the fruit, (b) contacting the heating surface, and (c) over a linear mechanism sliding toward the dry ice chamber.

inserted in the flat prickly pear surface to study variations during cauterization. As well after keeping the fruits for nine and 15 days at ambient storage, 10 prickly pears were cut nearby the sealed surface and in the middle of the fruit to measure TSS changes. Fruits stored for nine and 15 days at ambient storage were cut nearby the sealed surface and in the middle of the fruit to measure TSS and acidity changes.

4. Mango physicochemical analyses

Mango cuticle is thin and does not resist the high thermal gradient required by cauterization operations. Therefore, thermal treatments have to be applied carefully, mainly in the mango fruit abscission-pedicel interface.

Average biochemical maturity properties of fruits at early harvest for Haden, Kent, and Keitt were analyzed. These properties include pH, total soluble solids (TSS, °Brix), ascorbic acid (mg.100 g^{-1}), moisture content (%), and dry matter content DM (%). Kent and Keitt late varieties were harvested 137 and 148 days after fruit set, respectively. These results are similar with those obtained at Ghana plantations [93]. Mango trees with higher fertilization delayed fruit firmness decay. At the moment of harvest, fruits were green and firm for all varieties and fertilization regimes. After nine days of storage at 25°C, firmness decreased to 16.93 N for Kent fruits and remained firmer for Keitt mangoes. Chemical composition changes result from physiological and biochemical events controlled during fruit ripening [94]. Pectins are responsible for fruit texture and rise in the fifth week of mango fruit setting until the stone is formed. Pectins are responsible for fruit texture and rise five weeks after mango fruit setting until the stone is formed. Afterward, pectin content decreases, and fruit starts softening due to enzymatic degradation [64].

Fruits were harvested at a very green stage showing low TSS, acidity, and pH values (**Table 1**). As fruits mature nine days after, firmness decreased to 25.73, 16.93, and 32.91 N for Haden, Kent, and Keitt fruit, respectively (**Table 2**). After mango harvest, quality losses occur, affecting the content of nutritional components at different points during the handling chain [65].

Kent mangoes show a rapid decrease in firmness during ripening [95]. Kent mango trees with normal fertilization level produce fruits with high respiratory activity, lower ascorbic acid concentration, and fruit firmness drop [95].

Variety	Pulp pH	TSS (°Brix)	TA (% citric acid)	DM (%)	Firmness N
Haden	3.81	9.72	2.11	16.27	113.27
Kent	3.98	6.42	1.45	17.84	122.42
Keitt	3.66	7.63	2.43	17.85	121.05

Table 1.
Physicochemical analyses of different mango varieties considering pulp pH, TSS (total soluble solids), TA (Titratable acidity), DM (dry matter), and firmness of just harvested fruit.

Variety	Pulp pH	TSS (°Brix)	TA (% citric acid)	DM (%)	Firmness N
		18.32/	0.24/	19.20/	25.73/
Haden	5.12	17.56*	0.33*	18.86*	32.42*
		17.98/	0.21/	18.96/	16.93/
Kent	4.43	17.18*	0.31*	18.09*	22.42*
		15.72/	0.18/	18.55/	32.91/
Keitt	5.67	15.03*	0.27*	17.96	35.72*

**Measurements of fruits without latex removal.*

Table 2.
Physicochemical analyses of different mango varieties considering pulp pH, TSS (total soluble solids), TA (Titratable acidity), DM (dry matter), and firmness in the market place.

Lower content of potassium within tissues is related to higher acidity, while lower pulp pH responds to the fertilization regime [96]. Keitt mangoes showed the lower quantity of total soluble solids (15.72°Brix) and a low acidity of 0.18 (**Table 2**). On the other hand, Ca applications increased citric acid content in "Haden" mango fruits [97]; meanwhile pulp pH jumped to 5.12. Keitt mango showed higher vitamin C content than Kent and Haden fruits in their ripe phases, because of the inhibition of polyphenol oxidase (PPO). This mango variety provides better color and flavor retention during processing [98]. Mango refrigerated at 4°C tends to maintain the same TSS and TA during nine days of storage (**Figure 7a** and **b**). If the pedicel gets cauterized, mango TSS drops. Titratable acidity (**Figure 7b**) was significantly affected by fruit respiration, consuming organic acid.

4.1 Mango latex and diseases

Fruit fly control and removal of surface fungal diseases can be carried out by hot water immersion [99] and by hot air application. Hot water immersion is relatively easy to use, economic, and can provide accurate monitoring of fruit and water temperature. Mango fruits immersed in hot water at 52°C for 5 min eliminated anthracnose fungal infection [60]. Anthracnose infestation was not present after storing the fruit for 15 days at ambient temperature [100]. The effect of hot water treatment on pulp TSS was insignificant and mango visual quality remained outstanding. If green mature fruits are dipped for 20 min in water heated to 53°C, it will control both anthracnose and SER. If water is heated below 52°C, it is not effective to control anthracnose, and at 5 degrees warmer, it will scald the peel [101]. Hot water immersion without waxing affects the natural wax layer of the fruit surface, enhancing its senescence. Fruits coated with wax delay the ripening and extend their shelf life [102]. Keitt and Tommy Atkins mango fruits develop yellow pigments in the skin after hot water immersion [60]. TSS content of fruits immersed in hot water increased to 20°Brix, meanwhile untreated fruits remained at 17° Brix.

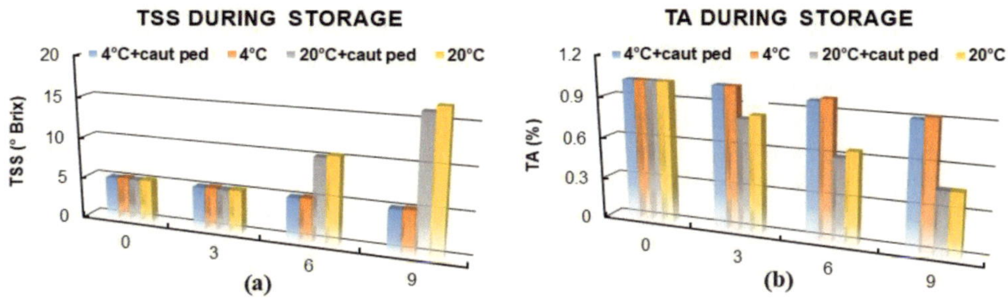

Figure 7.
Keitt mango (a) total soluble solid (TSS) concentration, and (b) Titratable acidity (TA) during the nine days of storage at 4 and 20°C with and without pedicel cauterization.

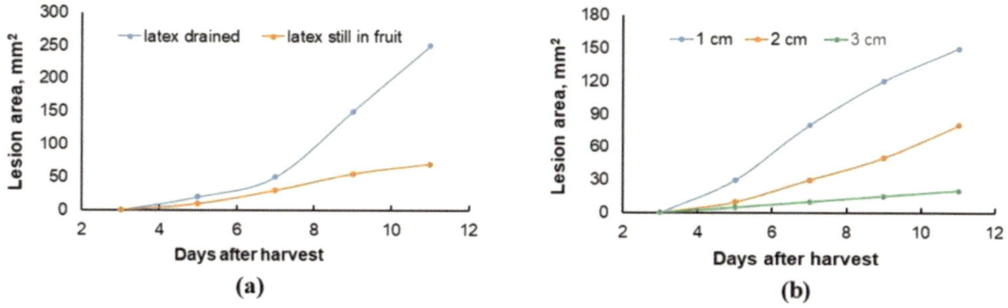

Figure 8.
Keitt mango anthracnose lesion area after several days of harvest (a) with and without latex, and (b) after petiole trimming.

In mangoes infected with SER, immersed in hot water and stored for 13 days, TSS content reached 19°Brix; fruits remained in 14°Brix if they were untreated [103].

At immature stage, anthracnose is not perceived, and the infection appears when mango ripens. Mango latex contains antifungal resorcinols and chitinase, so its retention during harvest will protect fruits against anthracnose and stem end rot [67]. Stem trimming deposits latex stains on the fruit surface, as pressurized sap stored in mango ducts falls on the fruit peel [65, 104]. Keitt mango fruit that preserved latex at harvest developed slightly smaller anthracnose lesions than fruits in which latex was drained (**Figure 8**). Keitt mango lesion area increases to 200 mm^2 after 10 days when fruits do not have latex (**Figure 8a**). Mango lesion

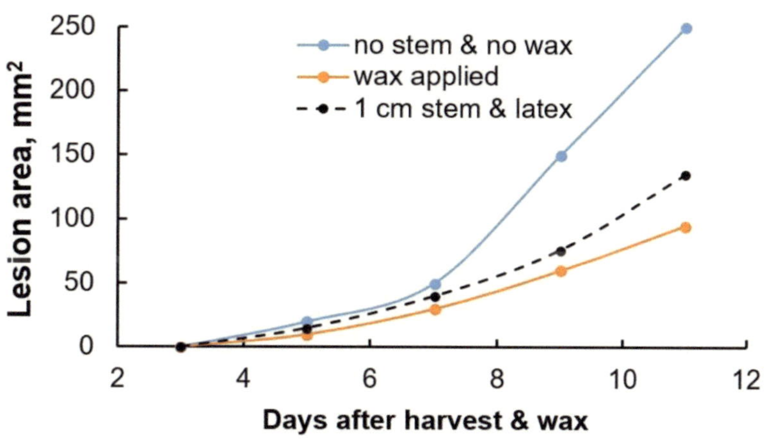

Figure 9.
Anthracnose lesion area several days of harvest for fruits cauterized and for mangoes after wax application.

corresponds to the black spot area growing on the fruit peel. When latex is present, the lesion only increases to 50 mm^2. The size of the remaining stem is correlated to the lesion area (**Figure 8a**). As it is longer and cauterized, less sap leaves the fruit, and it is more protected against pathogen infections. Higher anthracnose infection was noted in Keitt trees when more nitrogen was applied during fruit development [105]. This result was also found after analyzing "Willard" mango fruits [34].

When Keitt mango fruit stems were cauterized or their peduncle orifice covered with wax just after harvest, latex fluid remained within the fruit. Average anthracnose lesion was 38 and 54% smaller for wax and cauterization treatments, respectively, with respect to the control treatment after 11 days (**Figure 9**); no stem, wax, and latex were present on control fruits.

5. Prickly pear grippers and deformation experiments

Gripper suction cups grasp products by means of pressure difference [106, 107]. These grippers can be joined with other mechanisms easily, but are impractical for high-temperature grasping [108]. Modern granular-material grippers align themselves in malleable shapes to grasp the end product [108–110]. The prickly pear gripper used a grasping force of 40 N with a holding time of 30 s. The cauterizer robot (**Figure 6a**) presents a gripper moved by a mechanism containing two DC motors. One of the gripper fingers´ remains static during grasping, meanwhile the opposite finger presses the fruit; this finger moves using a DC motor. The second prototype used a pneumatic actuator. The slide actuator (**Figure 6b**) transports the six-finger gripper until a sensor detects its contact against the dry ice wall. A timer ensures that the fruit surface contacts the dry ice block during the right period. The pneumatic slider returns the fruit back to the pick and place area; this process takes 25 s. The end effector damaged the prickly pear during grasping and cauterization, when the fingers did not allow fruit movement. Fruit compression plotted in the vertical axis of **Figure 10** corresponds to the prickly pear deflection caused by finger pressing.

Prickly pears were sliced and cauterized by the robotic systems. Large prickly pears present an average diameter of 15 mm at the sliced section; smaller pears present a larger slice diameter ranging between 30 and 35 mm. Two clusters appear after plotting fruit firmness against pear compression (**Figure 10**). The black marks within the red circle show big fruits having firmness within 13 and 16 Ncm^{-2}. Fruit

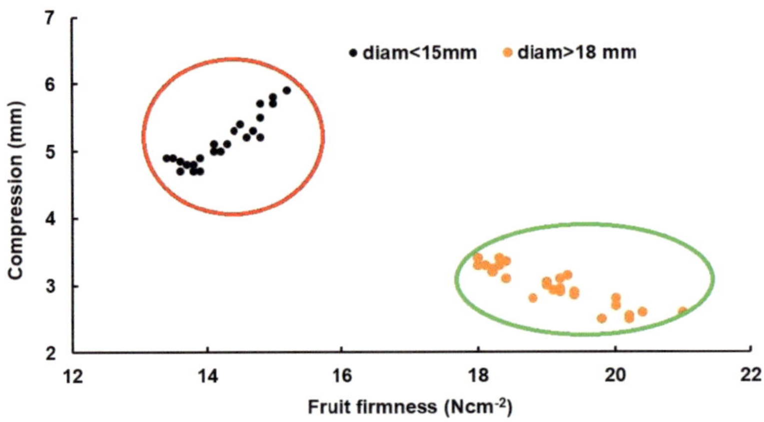

Figure 10.
Fruit firmness vs. compression for prickly pears having different slice diameter.

damage during processing decreased for pears compressed less than 3 mm. Orange markers show fruits with higher firmness (17.5–21.5 Ncm^{-2}), where the slicing area rises.

Prickly pear is a desert fruit with a thick peel. Pear firmness decreases once it is sliced (**Table 3**), and the fruit is destroyed when compression overpasses 5.2 mm. Red data in **Table 3** shows prickly pear values suffering some kind of damage. As the cauterizing diameter (ϱ) increases, fruit firmness drops and a lower pressure should be applied to avoid its destruction. Yellow fruits are softer and their tissue compresses easily. Therefore, yellow fruits are unable to withstand the cauterizing force (**Table 3**). As the prickly pear sliced area receives an orthogonal force, the airspaces within the pulp fill up. Pulp deformation takes place, growing sideways until the peel cannot withstand the pressure and explodes.

5.1 Prickly pear physicochemical analyses and measurements

Temperature measurements 2 mm within the pulp sliced area and at the middle of the prickly pear differ (**Figure 11**). The thermocouple placed 2 mm away from the sliced area reached only −4°C after 50 s, being hotter than the temperature of the dry ice block (−78°C). For the rotating robot (**Figure 6a**), fruit temperature decays after 50 s once the gripper contacts the dry ice surface, reaching its minimum temperature 10 s later. The green area in **Figure 11** shows negative pear temperature values in the sliced area during fruit cauterization contact. The complete temperature signal within the prickly pear during the cauterization cycle is shown in **Figure 11**. Fruit cauterization ended 125 s later, arriving to 17.4°C 145 s after; At this moment the slide system returned the pear back. Pulp temperature measurements acquired 15 mm below the sliced area were almost constant during the 6 min (**Figure 11**, dot line). Tissue temperature returns quicker to its natural thermal state (17.4°C) with the sliding system as shown by the red line, **Figure 11**. Cell walls have a more rigid contact when touching the dry ice chamber surface. Similar results were achieved by prickly pears that contacted the dry ice for 25 s.

TTS and total acidity (TA) were measured every 15 days after cutting three fruits at the center. TSS and TA monitoring was repeated in fruits stored for three months. Total soluble solids (TSS) concentration estimates the sugar content in the fruit and determines its degree of sweetness [111]. TSS concentration of prickly pears of cultivar "Blanca Cristalina" just after cryo-cauterization remained in 13.5°Brix. Measurements taken one, two, and three months later showed values of 13.4, 13.3, and 13.2°Brix, respectively. TSS minimum variations show that cryo-cauterization preserves fruit quality. Blanca Cristalina and Esmeralda fruits present 13.6 and

Diameter (mm)	Color	Firmness (Ncm^{-2})		Compression (mm)		Damage (%)
		Min	Max	Min	Max	
<15	green	16.12	16.82	2.5	3.2	0
15 < ϱ < 25	green	15.28	15.94	2.8	4.1	0
25 < ϱ < 35	green	14.47	15.35	4.2	5.5	50
<15	yellow	14.21	14.72	4.9	5.5	100
15 < ϱ < 25	yellow	13.73	14.15	5.1	5.5	100
25 < ϱ < 35	yellow	13.04	13.57	5.3	5.5	100

Table 3.
Green and yellow prickly pear firmness and compression having different slice diameters.

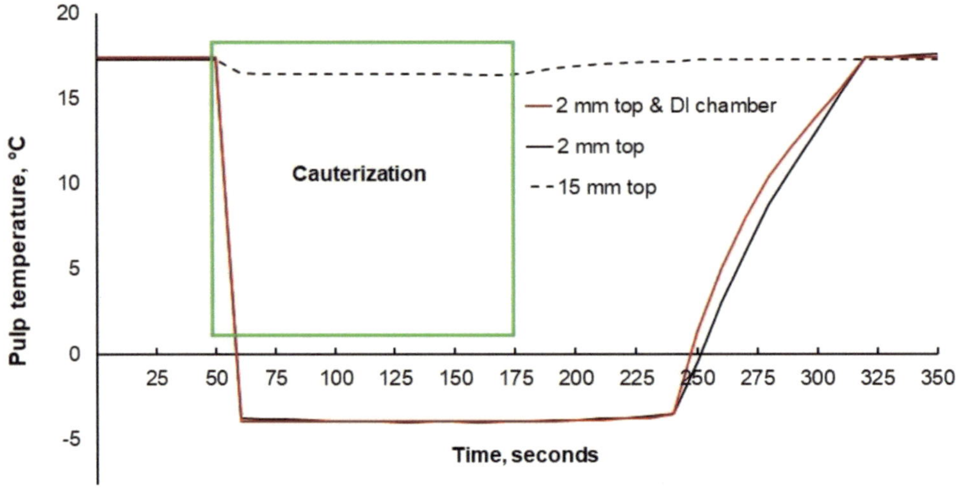

Figure 11.
Prickly pear pulp temperature monitored 2 mm and 15 mm away from the sliced surface during cauterization.

14°Brix at harvest, respectively [112]. Twenty-eight days later, TSS concentration was of 11.4 and 12° Brix for Blanca Cristalina and Esmeralda pears [112]. Cactus pears from the "Orito" cultivar presented 14.9°Brix after harvest and 14°Brix after 28 days later [111]. Blanca Cristalina TA values remained constant at 0.25% during the three months, so fruits remain acid and fruit acceptance high [111]. Blanca Cristalina and Esmeralda presented 0.27 and 0.29% of citric acid at harvest, respectively. After four weeks, it decreased to 0.18% in Blanca Cristalina [112]. For all the varieties, pulp citric acid decreased during ripening [113]. Although in these experiments cuticle thickness was not measured after heat treatments. Cuticle thickness reduction on some varieties was due to the effect of heat treatments [114]. The resistance provided by the cuticle against mechanical damage depends on the cuticle structure [115].

6. Conclusions

An increase in the quality and shelf life of mango fruit and prickly pear will increase their marketing worldwide. The first step to increase mango quality is to reduce fungal diseases such as anthracnose and stem end rot that appear due to environmental changes. Thermal treatments on mango fruits preserve their quality and reduce postharvest fruit disease infestation. Mango fruits must be harvested with care as mechanical damage of the stem end can start rotting in the fruit. Latex de-sapping after field harvest will reduce fruit sap burn.

Mango latex that contains antifungal resorcinols and chitinase should remain within the fruit to decrease anthracnose and stem end rot infestation. Stem channel thickness where latex flows can decrease after cauterization or by applying liquid paraffin. Two systems were developed to maintain latex after harvesting.

In the first system, a gripper grabbed the mango fruit and proceeded to cut the stem by means of two hot knifes maintained at 45°C. The cauterized pedicel presented burnt cells at the surface and reduced in size toward the stem end. This technique decreased anthracnose infestation by 50% after 11 days of storage when compared with de-sapped mango fruits. TSS concentration drops after pedicel cauterization. In the second equipment, warm paraffin wax was applied by a conventional gun to mango fruits without the stem end. Average anthracnose lesion was 38% smaller for paraffin application after 11 storage days than in untreated infested mangoes.

Prickly pears are native fruits from Mexico that grow in arid zones and have very important nutritional properties. Cauterization increased prickly pear fruits' shelf life over two months. Hot and cold cauterizer equipment extended shelf life without pathogen damage as the treatment seals the fruit and avoids dehydration. Two grippers were developed to cryo-cauterize prickly pears as this system is more energy-efficient than hot cauterization. The first gripper uses two fingers to press the thick fruit peel without damaging it. In this robotic system, the biggest disadvantage is the reduced dry ice pad duration. Warm air moves around the dry ice pad and melts in 5 h, so it has to be replaced. The second robotic system was more efficient as the dry ice block was within a chamber isolated from the air. Dry ice lasted for more than one day. This system used a six-finger gripper that moved over a pneumatic actuator, cryo-cauterizing a pear every 25 s. When the gripper contacted the dry ice wall, the temperature inside the fruit 2 mm away from the fruit sliced area was of −4°C. The temperature was measured with a thermocouple inserted in the fruit. Another temperature measurement was taken inside the pear 15 mm away from the sliced zone and the colder temperature was of 16°C. Gripper grasping damaged yellow fruits and its compression should be limited to 3 mm in green fruits. TSS and TA remained constant in cryo-cauterized fruit during the three months of fruit storage.

Acknowledgements

This work was supported by mango and prickly pear producers in Mexico by providing fruits for analysis. Also I would like to thank the DGIP, who funded the cauterizer project 21013-DTT. I appreciate the assistance of MI Angel Hernandez Facundo in the preparation of the drawings and the Language Department of the Universidad Autonoma Chapingo for reviewing this manuscript. As well I acknowledge the help from Biol and Ruth Perez in the analysis of data.

Conflict of interest

The authors declare no conflict of interest.

Author details

Federico Félix Hahn Schlam
Irrigation Department, Universidad Autonoma Chapingo, Texcoco, México

*Address all correspondence to: fhahn@correo.chapingo.mx

References

[1] Santos SF, Cardoso R, Borges ÍM, Almeida AC, Andrade ES, Ferreira IO, et al. Post-harvest losses of fruits and vegetables in supply centers in Salvador, Brazil: Analysis of determinants, volumes and reduction strategies. Waste Management. 2019;**101**:161-170. DOI: 10.1016/j.wasman.2019.10.007

[2] Rajapaksha L, Gunathilake DM, Pathirana SM, Fernando TN. Reducing post-harvest losses in fruits and vegetables for ensuring food security—Case of Sri Lanka. MOJ Food Processing and Technology. 2021;**9**(1):7-16. DOI: 10.15406/mojfpt.2021.09.00255

[3] Losada HR, Vieyra JE, Luna L, Cortés J, Vargas JM. Economic indicators, capacity of the ecosystem of prickly pear cactus (opuntia megacantha) and environmental services in Teotihuacan, México to Supply Urban Consumption. Journal of Agriculture and Environmental Sciences. 2017;**6**:85-91. DOI: 10.15640/jaes.v6n1a9

[4] Hernández-Guerrero S, Balois-Morales R, Hermosillo YA, Lopez G, Berumen-Varela G, Bautista-Rosales P, et al. Novel edible coating of starch-based stenospermocarpic mango prolongs the shelf life of mango "Ataulfo" fruit. Journal of Food Quality. 2020. pp. 9. Article ID 1320357. DOI: 10.1155/2020/1320357

[5] Rehana NS, Mansha N, Khaskheli MA, Khanzada LAM. Chemical control of stem end rot of mango caused by Lasiodiplodia theobromae. Pakistan Journal of Phytopathology. 2014;**26**:201-206

[6] Galsurker O, Diskin S, Duanis-Assaf D, Doron-Faigenboim A, Maurer D, Feygenberg O, et al. Harvesting mango fruit with a short stem-end altered endophytic microbiome and reduce stem-end rot. Microorganisms. 2020;**8**(4):558. DOI: 10.3390/microorganisms8040558

[7] Galsurker O, Diskin S, Maurer D, Feygenberg O, Alkan N. Fruit stem-end rot. Horticulturae. 2018;**4**(4):50. DOI: 10.3390/horticulturae4040050

[8] Iglesias DI, Cercós M, Colmenero-Flores JM, Naranjo MA, Ríos G, Carrera E, et al. Physiology of citrus fruiting. Brazilian Journal of Plant Physiology. 2007;**19**(4):333-362. DOI: 10.1590/S1677-04202007000400006

[9] Riederer M. Introduction: Biology of the plant cuticle. In: Riederer M, Müller C, editors. Biology of the Plant Cuticle. Hoboken, NJ: Blackwell Publishing; 2006. pp. 1-10. DOI: 10.1002/9780470988718.ch1

[10] Khanal BP, Knoche M. Mechanical properties of cuticles and their primary determinants. Journal of Experimental Botany. 2017;**68**(19):5351-5367. DOI: 10.1093/jxb/erx265

[11] Lai X, Khanal BP, Knoche M. Mismatch between cuticle deposition and area expansion in fruit skins allows potentially catastrophic buildup of elastic strain. Planta. 2016;**244**:1145-1156. DOI: 10.1007/s00425-016-2572-9

[12] Martin LBB, Rose JKC. There's more than one way to skin a fruit: Formation and functions of fruit cuticles. Journal of Experimental Botany. 2014;**65**(16):4639-4651. DOI: 10.1093/jxb/eru301

[13] Trivedi P, Nguyen N, Hykkerud AL, Häggman H, Martinussen I, Jaakola L, et al. Developmental and environmental regulation of cuticular wax biosynthesis in fleshy fruits. Frontiers in Plant Science. 2019;**10**:431. DOI: 10.3389/fpls.2019.00431

[14] Léchaudel M, Lopez-Lauri F, Vidal V, Sallanon H, Joas J. Response of the physiological parameters of mango

fruit (transpiration, water relations and antioxidant system) to its light and temperature environment. Journal of Plant Physiology. 2013;**170**:567-576. DOI: 10.1016/j.jplph.2012.11.009

[15] Tafolla-Arellano JC, Zheng Y, Sun H, Jiao C, Ruiz-May E, Hernández-Oñate MA, et al. Transcriptome analysis of mango (Mangifera indica L.) fruit epidermal peel to identify putative cuticle-associated genes. Scientific Reports. 2017;7:46163. DOI: 10.1038/srep46163

[16] Saladié M, Matas AJ, Isaacson T, Jenks MA, Goodwin SM, Niklas KJ, et al. A reevaluation of the key factors that influence tomato fruit softening and integrity. Plant Physiology. 2007; **144**:1012-1028. DOI: 10.1104/pp.107.097477

[17] Marques JPR, Spósito MB, Mello AFS, Amorim L, Mondin M, Appezzato-da-Glória B. Histopathology of black spot symptoms in sweet oranges. European Journal of Plant Pathology. 2012;**133**:439-448. DOI: 10.1007/s10658-011-9917-9

[18] Li J, Guo Y, Li Z, Lin Y, Liu L, Zhang X, et al. Super critical carbon dioxide and hexane extraction of wax from apple peel pomace: Content, composition, and thermal properties. Separation Science Technology. 2015;**50**:2230-2237. DOI: 10.1080/01496395.2015.1020951

[19] Solovchenko A, Merzlyak M. Optical properties and contribution of cuticle to UV protection in plants: Experiments with apple fruit. Photochemistry and Photobiology Science. 2003;**2**:861-866. DOI: 10.1039/B302478D

[20] Lara I, Belge B, Goulao L. The fruit cuticle as a modulator of postharvest quality. Postharvest Biology and Technology. 2014;**87**:103-112. DOI: 10.1016/j.postharvbio.2013.08.012

[21] Measham PF, Bound SA, Gracie AJ, Wilson SJ. Incidence and type of cracking in sweet cherry (Prunus avium L.) are affected by genotype and season. Crop and Pasture Science. 2009;**60**:1002-1008. DOI: 10.1071/CP08410

[22] Winkler A, Peschel S, Kohrs K, Knoche MJ. Rain cracking in sweet cherries is not due to excess water uptake but to localized skin phenomena. American Society for Horticultural Science. 2016;**141**(6):653-660. DOI: 10.21273/JASHS03937-16

[23] Atkinson RG, Brummell DA, Burdon JN, Patterson KJ, Schaffer RJ. Fruit growth, ripening and post-harvest. In: Brummell D, editor. Plants in Action. 2nd ed. Canberra, Australia: Australian Society of Plant Scientists; 2012. pp. 1-50

[24] Goulao LF, Oliveira CM. Cell wall modifications during fruit ripening: When a fruit is not the fruit. Trends in Food Science and Technology. 2008;**19**(1):4-25. DOI: 10.1016/j.tifs.2007.07.002

[25] Camacho-Vázquez C, Ruiz-Maya E, Guerrero-Analcoa JA, Elizalde-Contreras JM, Enciso-Ortiz EJ, Rosas-Saito G, et al. Filling gaps in our knowledge on the cuticle of mangoes (Mangifera indica) by analyzing six fruit cultivars: Architecture/structure, postharvest physiology and possible resistance to fruit fly (Tephritidae) attack. Postharvest Biology and Technology. 2019;**148**:83-96. DOI: 10.1016/j.postharvbio.2018.10.006

[26] Lara I, Heredia A, Domínguez E. Shelf life potential and the fruit cuticle: The unexpected player. Frontiers in Plant Science. 2019;**10**:770. DOI: 10.3389/fpls.2019.00770

[27] Paniagua C, Posé S, Morris VJ, Kirby AR, Quesada MA, Mercado JA. Fruit softening and pectin disassembly: An overview of nanostructural pectin

modifications assessed by atomic force microscopy. Annals of Botany. 2014;**114**(6):1375-1383. DOI: 10.1093/aob/mcu149

[28] Posé S, Paniagua C, Matas AJ, Gunning AP, Morris VJ, Quesada MA, et al. A nanostructural view of the cell wall disassembly process during fruit ripening and postharvest storage by atomic force microscopy. Trends in Food Science and Technology. 2019;**87**:47-58. DOI: 10.1016/j.tifs.2018.02.011

[29] Mercado JA, Pliego-Alfaro F, Quesada MA. Fruit shelf life and potential for its genetic improvement. In: Jenks MA, Bebeli PJ, editors. Breeding for Fruit Quality. Hoboken, NJ: Wiley; 2011. pp. 81-104. DOI: 10.1002/9780470959350

[30] Zhang L, Chen F, Yang H, Sun X, Liu H, Gong X, et al. Changes in firmness, pectin content and nanostructure of two crisp peach cultivars after storage. Food Science and Technology. 2010;**43**(1):26-32. DOI: 10.1016/j.lwt.2009.06.015

[31] Liu H, Chen F, Lai S, Tao J, Yang H, Jiao Z. Effects of calcium treatment and low temperature storage on cell wall polysaccharide nanostructures and quality of postharvest apricot (Prunus armeniaca). Food Chemistry. 2017;**15**(225):87-97. DOI: 10.1016/j.foodchem.2017.01.008

[32] San A, Hofman PJ, Joyce DC, Macnish AJ, Marques JR, Webb RI, et al. Diurnal harvest cycle and sap composition affect under-skin browning in 'Honey Gold' mango fruit. Frontiers in Plant Science. 2019;**10**:1093. DOI: 10.3389/fpls.2019.01093

[33] Hassan MK, Irving DE, Dann EK, Coates LM, Hofman PJ. Sap properties and alk(en)ylresorcinol concentrations in Australian-grown mangoes. Annals of Applied Biology. 2009;**154**:419-427. DOI: 10.1111/j.1744-7348.2008. 00313.x

[34] Karunanayake C, Sinniah G, Adikaram N, Abayasekara C, Wijayasekara D. Retention of latex at harvest, enhanced mango (Mangifera indica L.) fruit resistance and reduced anthracnose and stem-end rot. Australasian Plant Pathology. 2015;**44**:113-119. DOI: 10.1007/s13313-014-0330-7

[35] Esguerra EB, Bautista OK. Quality and safety in agri-food chains in the Philippines: The case of mango. Acta Horticulturae. 2013;**989**:239-243. DOI: 10.17660/ActaHortic.2013.989.31

[36] Krishnapillai N, Wijeratnam S. Sap burn injury management of mangoes (Mangifera indica L.) in Sri Lanka. Pakistan Journal of Botany. 2016; **48**:2147-2152

[37] Krishnapillai N, Wijeratnam W. Sap volatile components in relation to susceptibility of anthracnose and aspergillus rot of mangoes (Mangifera indica L.). The Journal of Horticultural Science and Biotechnology. 2017;**92**(2):206-213. DOI: 10.1080/14620316.2016.1249962

[38] Hahn F. Cauterizer technology increases cactus pear shelf life. In: Kahramanoglu I, editor. Postharvest Handling. Rijeka: IntechOpen; 2017. pp. 141-163. DOI: 10.5772/intechopen.68845

[39] Valero-Galván J, González-Fernández R, Sigala-Hernández A, Núñez-Gastélum JA, Ruiz-May E, Rodrigo-García J, et al. Sensory attributes, physicochemical and antioxidant characteristics, and protein profile of wild prickly pear fruits (O. macrocentra Engelm., O. phaeacantha Engelm., and O. engelmannii Salm-Dyck ex Engelmann.) and commercial prickly pear fruits (O. ficus-indica (L.) Mill.). Food Research International. 2021;**140**:109909. DOI: 10.1016/j.foodres.2020.109909

[40] Illoldi-Rangel P, Ciarleglio M, Sheinvar L, Linaje M,

Sanchez-Cordero SS. Opuntia in Mexico: Identifying priority areas for conserving biodiversity in a multi-use landscape. PLoS One. 2012;**7**(5):1-16. DOI: 10.1371/journal.pone.0036650

[41] Bourhia M, Elmahdaoui H, Ullah R, Bari A, Benbacer L. Promising physical, physicochemical, and biochemical background contained in peels of prickly pear fruit growing under hard ecological conditions in the mediterranean countries. BioMed Research International. 2019. 8 pages. Article ID: 9873146. DOI: 10.1155/2019/9873146

[42] Bourhia M, Elmahdaoui H, Ullah R, Ibenmoussa S, Shahat A. Physicochemical evaluation of the fruit pulp of Opuntia spp growing in the Mediterranean area under hard climate conditions. Open Chemistry. 2020;**18**(1):565-575. DOI: 10.1515/chem-2020-0097

[43] Stintzing FC, Herbach KM, Mosshammer MR, Carle R, Yi W, Sellappan S, et al. Color, betalain pattern, and antioxidant properties of cactus pear (Opuntiaspp.) clones. Journal of Agricultural and Food Chemistry. 2005;**53**(2):442-451. DOI: 10.1021/jf048751y

[44] Hahn-Schlam F, Valle-Guadarrama S, Jenkins T. Robotic cactus pear cryocauterization increases storage life. Postharvest Biology and Technology. 2019;**147**:132-138. DOI: 10.1016/j.postharvbio.2018.09.014

[45] Añorve MJ, Aquino BEN, Mercado SE. Effect of controlled atmosphere on the preservation of cactus pears. Acta Horticulturae. 2006;**728**:211-216. DOI: 10.17660/ActaHortic.2006.728.30

[46] Piga A, Del Caro A, Pinna I, Agabbio M. Changes in ascorbic acid, polyphenol content and antioxidant activity in minimally processed cactus pear fruits. LWT – Food Science and Technology. 2003;**36**:257-262. DOI: 10.1016/S0023-6438(02)00227-X

[47] Raval AH, Solanki SC, Yadav R. A simplified heat transfer model for predicting temperature change inside food package kept in cold room. Journal of Food Science and Technology. 2013;**50**(2):257-265.DOI:10.1007/s13197-011-0342-z

[48] Ambaw A, Fadiji T, Opara UL. Thermo-mechanical analysis in the fresh fruit cold chain: A review on recent advances. Food. 2021;**10**(6):1357. DOI: 10.3390/foods10061357

[49] Brizzolara S, Manganaris GA, Fotopoulos V, Watkins CB, Tonutti P. Primary metabolism in fresh fruits during storage. Frontiers in Plant Science. 2020;**11**:1-16. DOI: 10.3389/fpls.2020.00080

[50] Arah I, Ahorbo G, Anku E, Kumah E, Amaglo H. Postharvest handling practices and treatment methods for tomato handlers in developing countries: A mini review. Advances in Agriculture. Volume 2016. 8 pages. Article ID: 6436945. DOI: 10.1155/2016/6436945

[51] Akbudak B, Akbudak N, Seniz V, Eris A. Effect of pre-harvest harpin and modified atmosphere packaging on quality of cherry tomato cultivars "Alona" and "Cluster". British Food Journal. 2012;**114**(2):180-196. DOI: 10.1108/00070701211202377

[52] Saltveit ME. Water Loss from harvested horticultural commodities. In: Pareek S, editor. Postharvest Ripening Physiology of Crops. 1st ed. London: CRC Press; 2016. pp. 139-156. DOI: 10.1201/b19043

[53] Uyar R, Erdoğdu F. Numerical evaluation of spherical geometry approximation for heating and cooling of irregular shaped food products.

Journal of Food Science. 2012;**77**:E166-E175. DOI: 10.1111/j.1750-3841.2012.02769.x

[54] Cinquanta L, Di Matteo M, Estia M. Physical pre-treatment of plums (Prunus domestica). Part 2. Effect on the quality characteristics of different prune cultivars. Food Chemistry. 2002;**79**(2):233-238. DOI: 10.1016/S0308-8146(02)00138-3

[55] Wang L, Sun D. Recent developments in numerical modelling of heating and cooling processes in the food industry—A review. Trends in Food Science and Technology. 2003;**14**(10):408-423. DOI: 10.1016/S0924-2244(03)00151-1

[56] Sui Y, Wisniewski M, Droby S, Norelli J, Liu J. Recent advances and current status of the use of heat treatments in postharvest disease management systems: Is it time to turn up the heat? Trends in Food Science and Technology. 2016;**51**:34-40. DOI: 10.1016/j.tifs.2016.03.004

[57] Tahir II, Johansson E, Olsson ME. Improvement of apple quality and storability by a combination of heat treatment and controlled atmosphere storage. HortScience. 2009;**44**:1648-1654. DOI: 10.21273/HORTSCI.44.6.1648

[58] Queb-González DB, Lopez-Malo A, Sosa-Morales ME, Villa-Rojas R. Postharvest heat treatments to inhibit Penicillium digitatum growth and maintain quality of Mandarin (Citrus reticulata blanco). Heliyon. 2020;**6**(1):e03166. DOI: 10.1016/j.heliyon.2020.e03166

[59] Ibarz A, Barbosa-Canovas GV. Unit Operations in Food Engineering. 1st ed. London: CRC Press; 2002. p. 920. DOI: 10.1201/9781420012620

[60] Jacobi K, Macrae E, Hetherington S. Postharvest heat disinfestation treatment of mango fruit. Scientia Horticulturae. 2001;**89**:171-193. DOI: 10.1016/S0304-4238(00)00240-5

[61] Shellie K, Mangan R. Cooling method and fruit weight: Efficacy of hot water quarantine treatment for control of mexican fruit fly in mango. HortScience. 2002;**37**(6):910-913. DOI: 10.21273/HORTSCI.37.6.910

[62] Yimyong S, Datsenka T, Handa A, Seraypheap K. Hot water treatment delays ripening-associated metabolic shift in 'Okrong' mango fruit during storage. Journal of the American Society for Horticultural Science. 2011;**136**:441-451. DOI: 10.21273/JASHS.136.6.441

[63] Tharanathan RN, Yashoda HM, Prabha TN. Mango (Mangifera indica L.), "The King of Fruits"—An overview. Food Reviews International. 2006;**22**(2):95-123. DOI: 10.1080/87559120600574493

[64] Maldonado-Celis ME, Yahia Elhadi M, Bedoya R, Landázuri P, Loango N, Aguillón J, et al. Chemical composition of mango (Mangifera indica L.) fruit: Nutritional and phytochemical compounds. Frontiers in Plant Science. 2019;**10**:1073. DOI: 10.3389/fpls.2019.01073

[65] Esguerra EB, Rolle R. Post-Harvest Management of Mango for Quality and Safety Assurance. Guidance for Horticultural Supply Chain Stakeholders. Rome: Food and Agriculture Organization of the United Nations; 2018. pp. 1-24

[66] Kangatharalingam N, Pierce ML, Bayles MB, Essenberg M. Epidermal anthocyanin production as an indicator of bacterial blight resistance in cotton. Physiological and Molecular Plant Pathology. 2002;**61**:189-195. DOI: 10.1006/pmpp.2002.0434

[67] Karunanayake KOLC, Sinniah GD, Adikaram NKB, Abayasekara CL. Cultivar differences in antifungal

activity and the resistance to postharvest anthracnose and stem-end rot in mango (Mangifera indica L.). Australasian Plant Pathology. 2014;**43**:151-159. DOI: 10.1007/s13313-013-0257-4

[68] Secretaria L, Bayogan ER, Lubaton CD, Majomot AMC, Ekman J, Goldwater A. Effect of harvest time, delay in destemming and desapping treatment on the sap volume and visual quality of 'Carabao' mango fruit. Walailak Journal of Science and Technology (WJST). 2021;**18**:7. DOI: 10.48048/wjst.2021.9076

[69] Higuchi H, Sakuratani T. Water dynamics in mango (Mangifera indica L.) fruit during the young and mature fruit seasons as measured by the stem heat balance method. Journal of the Japanese Society for Horticultural Science. 2015;**75**:11-19. DOI: 10.2503/jjshs.75.11

[70] Kamle M, Kumar P, Gupta VK, Tiwari AK, Misra AK, Pandey BK. Identification and phylogenetic correlation among colletotrichum gloeosporioides pathogen of anthracnose for mango. Biocatalysis and Agricultural Biotechnology. 2013;**2**:285-287. DOI: 10.1016/j.bcab.2013.04.001

[71] Alkan N, Fortes AM. Insights into molecular and metabolic events associated with fruit response to post-harvest fungal pathogens. Frontiers in Plant Science. 2015;**6**:889. DOI: 10.3389/fpls.2015.00889

[72] Navas E, Fernández R, Sepúlveda D, Armada M, Gonzalez-de-Santos P. Soft grippers for automatic crop harvesting: A review. Sensors. 2021;**21**:2689. DOI: 10.3390/s21082689

[73] Blanes C, Mellado M, Ortiz C, Valera A. Technologies for robot grippers in pick and place operations for fresh fruits and vegetables. Spanish Journal of Agricultural Research. 2011;**9**:1130-1141. DOI: 10.5424/sjar/20110904-501-10

[74] van Henten E, Hemming J, van Tuijl B, Kornet JG, Meuleman J, Bontsema J, et al. An autonomous robot for harvesting cucumbers in greenhouses. Autonomous Robots. 2002;**13**:241-258. DOI: 10.1023/A:1020568125418

[75] Jia B, Zhu A, Yang SX, Mittal GS. Integrated gripper and cutter in a mobile robotic system for harvesting greenhouse products. In: Proceedings of the 2009 IEEE International Conference on Robotics and Biomimetics (ROBIO); 19-23 December 2009; New York: IEEE; 2009:1778-1783. DOI: 10.1109/ROBIO.2009.5420430

[76] Ceccarelli M, Figliolini G, Ottaviano E, Mata A, Criado E. Designing a robotic gripper for harvesting horticulture products. Robotica. 2000;**18**(1):105-111. DOI: 10.1017/S026357479900226X

[77] Bachche S, Oka K. Performance testing of thermal cutting systems for sweet pepper harvesting robot in greenhouse horticulture. Journal of System Design and Dynamics. 2013;**7**:36-51. DOI: 10.1299/jsdd.7.36

[78] Liu J, Hu Y, Xu X, Li P. Feasibility and influencing factors of laser cutting of tomato peduncles for robotic harvesting. African Journal of Biotechnology. 2011;**10**(69):15552-15563. DOI: 10.5897/AJB.9000253

[79] Blanes C, Cortés López V, Ortiz Sánchez MC, Mellado Areche M, Talens OP. Non-destructive assessment of mango firmness and ripeness using a robotic gripper. Food and Bioprocess Technology. 2015;**8**(9):1914-1924. DOI: 10.1007/s11947-015-1548-2

[80] Cortés V, Blanes C, Blasco J, Ortiz C, Aleixos N, Mellado M, et al. Integration

of simultaneous tactile sensing and reflectance visible and near-infrared spectroscopy in a robot gripper for mango quality assessment. Biosystems Engineering. 2017;**166**:112-123. DOI: 10.1016/j.biosystemseng.2017.08.005

[81] Pardo-Alonso J-L, Carreño-Ortega Á, Martínez-Gaitán C-C, Golasi I, Gómez GM. Conventional industrial robotics applied to the process of tomato grafting using the splicing technique. Agronomy. 2019;**9**(12):880. DOI: 10.3390/agronomy9120880

[82] Karunanayake C. A qualitative analysis of mango (Mangifera indica L.) latex and anatomy of latex canals. Journal of Science. 2019;**10**(2):11-20. DOI: 10.4038/jsc.v10i2.21

[83] Kahramanoğlu İ, Usanmaz S, Okatan V, Wan C. Preserving postharvest storage quality of fresh-cut cactus pears by using different bio-materials. CABI Agriculture and Bioscience. 2020;**1**:1-13. DOI: 10.1186/s43170-020-00008-5

[84] Cefola M, Renna M, Pace B. Marketability of ready-to-eat cactus pear as affected by temperature and modified atmosphere. Journal of Food Science and Technology. 2014;**51**(1):25-33. DOI: 10.1007/s13197-011-0470-5

[85] Harker FR, Hallett IC, White A, Seal AG. Measurement of fruit peelability in the genus actinidia. Journal of Texture Studies. 2011;**42**(4):237-246. DOI: 10.1111/j.1745-4603.2010.00270.x

[86] Hahn F. Equipo industrial para la cauterización de tunas. IMPI Patent granted on Nov 16, 2016. Patent No: MX 343799 B. Pp 42

[87] Hahn F. Proceso y equipo industrial para cauterizar la superficie afectada por el corte en tunas, después de su cosecha. [Process and industrial equipment to cauterize prickle pear sliced surface after harvest]. IMPI Patent granted on Jan 22, 2018. Patent No: MX 353650 B. Pp 44

[88] Hahn F, Cruz J, Barrientos A, Perez R, Valle S. Optimal pressure and temperature parameters for prickly pear cauterization and infrared imaging detection for proper sealing. Journal of Food Engineering. 2016;**191**:131-138. DOI: 10.1016/j.jfoodeng. 2016.07.013

[89] Gheribi R, Khwaldia K. Cactus mucilage for food packaging applications. Coatings. 2019;**9**(10):655. DOI: 10.3390/coatings9100655

[90] Fox DI, Pichler T, Yeh DH, Alcantar NA. Removing heavy metals in water: The interaction of cactus mucilage and arsenate (As (V)). Environmental Science & Technology. 2012;**46**(8):4553-4559. DOI: 10.1021/es2021999

[91] Durán-García H, Guarneros-García O, Jiménez Delgado C, Rossel-Kipping E, Pulido-Delgado J. Structural design of a mechanical arm for harvest of cactus pear type Alfajayucan. Journal of applied research and technology. 2016;**14**(2):140-147. DOI: 10.1016/j.jart.2016.04.002

[92] Hahn F, Hernandez A. Automated chamber for prickle pear cauterization. Advances in Robotics and Mechanical Engineering. 2021;**3**(1):230-235. DOI: 10.32474/ARME.2021.03.000152

[93] Moomin A, Dzarkwei AL, Kobla AN. Relation of harvesting time on physicochemical properties of Haden, Kent, Palmer and Keitt mango varieties for export and local markets. Journal of Horticulture and Postharvest Research. 2021;**4**(1):87-100. DOI: 10.22077/jhpr.2020.3170.1126

[94] Gill PPS, Jawandha SK, Kaur N, Singh N. Physicochemical changes

during progressive ripening of mango (Mangifera indica L.) cv. Dashehari under different temperature regimes. Journal of Food Science and Technology. 2017;**54**(7):1-7. DOI: 10.1007/s13197-017-2632-6

[95] García MR, López JA, Saucedo VC, Salazar GS, Suárez EJ. Maduración y calidad de frutos de mango 'Kent' con tres niveles de fertilización. Revista mexicana de ciencias agrícolas. 2015;**6**(4):665-678

[96] Lobit P, Genard M, Soing P, Habib R. Modelling malic acid accumulation in fruits: Relationships with organic acids, potassium, and temperature. Journal of Experimental Botany. 2006;**57**(6):1471-1483. DOI: 10.1093/jxb/erj128

[97] Romero-Gomezcaña N, Sanchez-Garcia P, Rodríguez-Alcázar JV, Crescenciano SV. Aplicación foliar de calcio y su relación con la calidad en frutos de mango cv. Haden. Agricultura técnica en México. 2006;**32**:5-15

[98] Palma-Orozco G, Marrufo-Hernandez NA, Sampedro JG, Najera H. Purification and partial biochemical characterization of polyphenol oxidase from mango (Mangifera indica cv. Manila). Journal of Agricultural and Food Chemistry. 2014;**62**:9832-9840. DOI: 10.1021/jf5029784

[99] López JM, Castaño ZJ. Management of mango anthracnose [Glomerella cingulata (Stoneman) Spauld. & H. Schrenk] in post-harvest. Agronomía. 2010;**18**:47-57

[100] Angasu ON, Dessalgne OG, Tadesse TN. Effect of hot water treatment on quality and incidence of postharvest disease of mango (Mangifera indicia L.) fruits. Asian Journal of Plant Sciences. 2014;**13**:87-92. DOI: 10.3923/ajps.2014.87.92

[101] Alvindia DG, Acda MA. Revisiting the efficacy of hot water treatment in managing anthracnose and stem-end rot diseases of mango cv. 'Carabao'. Crop Protection. 2015;**67**:96-101. DOI: 10.1016/j.cropro.2014.09.016

[102] Govender V, Korsten L, Sivakumar D. Semi-commercial evaluation of Bacillus licheniformis to control mango postharvest diseases in South Africa. Postharvest Biology and Technology. 2005;**38**(1):57-65. DOI: 10.1016/j.postharvbio.2005.04.005

[103] Mon YY, Win NK, Aye SS, Soe YY, Naing TAA. Effect of hot water treatment on mango postharvest diseases: Stem end rot and anthracnose. Journal of Agricultural Research. 2017;**4**(2):79-85

[104] Loveys BR, Robinson SP, Brophy JJ, Chacko EK. Mango sapburn: Components of fruit sap and their role in causing skin damage. Australian Journal of Plant Physiology. 1992;**19**:449-457. DOI: 10.1071/PP9920449

[105] Bally ISE, Hofman PJ, Irving DE, Coates LM, Dann EK. The effects of nitrogen on postharvest disease in mango (mangifera indica l. 'Keitt'). Acta Horticulturae. 2009;**820**:365-370. DOI: 10.17660/ActaHortic.2009.820.42

[106] Sdahl M, Kuhlenkoetter B. CAGD-computer aided gripper design for a flexible gripping system. International Journal Advanced Robot System. 2006;**2**(2):135-138. DOI: 10.5772/5795

[107] Mantriota G. Optimal grasp of vacuum grippers with multiple suction cups. Mechanism and Machine Theory. 2007;**42**:18-33. DOI: 10.1016/j.mechmachtheory.2006.02.007

[108] Monkman G, Hesse S, Steinmann R, Schunk H. Robot Grippers. Weinheim, Germany: Wiley-VCH; 2007. p. 439

[109] Majmudar T, Sperl M, Luding S, Behringer R. Jamming transition in

granular systems. Physics Review Letters. 2007;**98**:058001. DOI: 10.1103/PhysRevLett.98.058001

[110] Corwin E, Jaeger H, Nagel S. Structural signature of jamming in granular media. Nature. 2005;**205**(435):1075-1078. DOI: 10.1038/nature03698

[111] Andreu-Coll L, García-Pastor ME, Valero D, Amorós A, Almansa MS, Legua P, et al. Influence of storage on physiological properties, chemical composition, and bioactive compounds on cactus pear fruit (Opuntia ficus-indica (L.) Mill.). Agriculture. 2021;**11**(1):62. DOI: 10.3390/agriculture11010062

[112] Hernández-Pérez T, Carrillo-López A, Guevara-Lara F, Cruz-Hernández A, Paredes-López O. Biochemical and nutritional characterization of three prickly pear species with different ripening behavior. Plant Foods for Human Nutrition. 2005;**60**(4):195-200. DOI: 10.1007/s11130-005-8618-y

[113] Saenz C, Estevez SE, Mecklenburg S. Cactus pear fruit: A new source for natural sweetener. Plant Foods for Human Nutrition. 1998;**52**:141-149. DOI: 10.1023/A:1008033704523

[114] Castañeda J, Corrales J, Terrazas T, Colinas M. Effect of vapor heat treatments on weight loss reduction and epicuticular changes in six varieties of cactus pear fruit (Opuntia spp.). Journal of the Professional Association for Cactus Development. 2010;**12**:37-47

[115] Knoche M, Beyer M, Peschel S, Oparlakov B, Bukovac MJ. Changes in strain and deposition of cuticle in developing sweet cherry fruit. Physiologia Plantarum. 2004;**120**:667-677. DOI: 10.1111/j.0031-9317.2004.0285.x

Chapter 5

Postharvest Diseases of Vegetable Crops and Their Management

Atma Nand Tripathi, Shailesh Kumar Tiwari and Tushar Kanti Behera

Abstract

Vegetable crops have an important role in food and nutrition and maintain the health of soil. India is the second-largest producer of vegetables in the world with a 16% (191.77 MT) share of global vegetable production. Every year, diseases cause postharvest losses (40–60%) in vegetable crops due to their perishable nature under field (15–20%), packaging and storage (15–20%), and transport (30–40%). Profiling, detection, and diagnosis of postharvest vegetable pathogens (diseases) are essential for better understanding of pathogen and formulation of safe management of postharvest spoilage of vegetables. The vegetable produce is spoiled by postharvest pathogens and makes them unfit for human consumption and market due to the production of mycotoxins and other potential human health risks. Genera of fungal pathogens viz. *Alternaria*, *Aschochyta, Colletotrichum, Didymella, Phoma, Phytophthora, Pythium, Rhizoctonia, Sclerotinia, Sclerotium,* and bacterial pathogens viz. *Erwinia* spp., *Pseudomonas* spp., *Ralstonia solanacearum*, *Xanthomonas euvesictoria* were recorded as postharvest pathogens on vegetable crops. Fruit rot incidence of several post-harvest pathogens viz. *Alternaria solani* (30%), *Phytophthora infestans* (15%), *Rhophitulus solani* (30%), *Sclerotium rolfsii* (30%) fruit rot and *X. euvesictoria* (5%) canker on tomato; *Colletotrichum dematium* fruit rot (20%) on chili; *Phomopsis vexans* (60%) fruit rot on brinjal was recorded. *Didymella* black rot and *Colletotrichum* anthracnose were recorded on fruits of bottle gourd, pumpkin, ash gourd, and watermelon. Important leguminous vegetable crops are infected by postharvest pathogens viz. *Ascochyta pisi, Colletotrichum lindemuthianum* (Anthracnose), *Sclerotinia sclerotiorum* (white rot) and *Pseudomonas syringae* pv. *phaseolicola* (blight), *Sclerotinia* white rot, *Alternaria* blight. However, *Xanthomonas* black rot (10%) on cabbage and *Pectinovora* (*Erwinia*) soft rot (19%) were recorded as emerging post-harvest pathogens on cauliflower.

Keywords: vegetable diseases, plant pathogens, diseases management, seed-borne, soil-borne diseases

1. Introduction

India is the second-largest producer of vegetables in the world after China, and shares about 16% of global vegetable production [1]. Processed vegetables have been exported at a compounded annual growth rate in the volume of 16% and in value of 25% [2, 3]. Vegetables have a significant role in enhancing farm income, sustainable global food as well as nutritional security. Vegetables suffer from several fungal and bacterial postharvest diseases [4–6]. Postharvest losses in vegetables are reported up to 30–40% owing to poor postharvest practices [7].

Fungicide is commonly applied for post-harvest disease control. Hot air, curing and hot-water brushing reduces disease incidence and increases the efficacy of antagonists. Biocontrol agents and botanicals may also reduce the amount of fungicide frequently used in postharvest disease management. Biocontrol of postharvest diseases of vegetable crops has great potential under storage conditions and biological products/biopesticides are available in the market. The biopesticides Ecogen US (Aspire™), Azotobactor (Bio-Save™), and Anchor (Yield Plus™) are involved to combine products with a low level of fungicide and salt solutions (calcium chloride or sodium bicarbonate @ 1–2%) and other food additives to improve efficacy against postharvest diseases. EcoSMART formulation based on rosemary oil, viz. EcoTrol™, Sporan™ (fungicide) and eugenol oil formulation Mataran™ (weedicides) are recognized as safe plant protectants. Therefore, the postharvest application of eco-friendly control methods may be exploited to manage the disease of vegetables.

2. Economical and health impact of postharvest diseases of vegetables

Postharvest diseases cause qualitative and quantitative losses of vegetables and make them unfit for human consumption due to potential health risks. A large number of postharvest diseases are caused by black, white, and yellow fungi-derived carcinogenic mycotoxins and mutagenic secondary metabolites [8]. Losses due to postharvest disease may occur during the handling of produce from harvest to consumption. Primary and secondary agricultural practices are also important and costs such as harvesting, packaging, and transport must be taken into account when estimating the value of the produce lost as a result of post-harvest wastage. Fresh vegetables are highly perishable, and they have relatively short shelf lives. Fresh vegetables are living, respiring tissues that start senescing immediately after harvest. They are mostly comprised of water, with most having 90–95% moisture content. Because of the perishable nature of vegetables, special skills are required for postharvest handling. *Aspergillus flavus* is a saprophytic soil inhabitant fungus that infects postharvest vegetables and produces carcinogenic secondary metabolite aflatoxin in tropical, subtropical, and temperate geographic regions of the world. It also causes animal and human diseases (causing aflatoxicosis and/or liver cancer) due to consumption of contaminated food and feed and through invasive growth (causing aspergillosis), which is often fatal to humans who are immunocompromised [9]. A holistic approach is needed for regulating aflatoxins under the trade/export market with biosecurity including bio-warfare, biodiversity, and biosafety for liberalized trade under the World Trade Organization (WTO) [10].

3. Challenges of postharvest losses in vegetable crops

Application of good postharvest management practices which are supported by good technologies and also improving postharvest systems will maintain the quality of vegetables and reduce quantitative losses. Losses in vegetables are the result of (i) poor knowledge about the right harvesting index; thus, a large proportion of the harvested beans are usually over-mature (ii) poor handling practices, such as the use of plastic sacks for bulk packaging and transportation which results in mechanical damage that serves as entry points for disease-causing organisms leading to rotting of the pods (iii) poor transport practices such as the use of trucks that have no cover, thus exposing the produce to direct sunlight and high temperature (iv) the absence of low-temperature storage facilities and transport systems, and (v) rough handling practices during distribution in retail markets.

4. Causes of postharvest diseases

In general, postharvest diseases and losses of vegetables are incited by fungi and bacteria. Postharvest diseases are often classified on the basis of the infection as "quiescent"or "latent", where the pathogen infects before harvest in the field. Examples of postharvest diseases arising from quiescent infections include anthracnose of various vegetables caused by *Colletotrichum* spp. and gray mold rot caused by *Botrytis cinerea*. Some pathogens infect vegetables after harvest during storage and transport, which is called postharvest infection e.g. nesting disease of pea caused by *Pythium* species or *Rhizopus* species. Microbes infect horticultural produce and spread rapidly due to a lack of natural defense mechanisms in the tissues of the produce. Management of postharvest spoilage is becoming a very difficult task because the pesticides/chemicals available are rapidly declining with consumer concern for food safety.

5. Detection of postharvest pathogens

Pathogens were isolated on agar medium and identified on the basis of macroscopic and microscopic analysis of colony and conidia/spore morphology by Microscopy, Sero-diagnostics (ELISA, Dot-blot assays), and nucleic acid (PCR) based methods.

Why do we need, want, or should detect emerging postharvest pathogens (diseases) in vegetable crops?

- Determine presence and quantity of the pathogen (s) for quarantine legislation.
- Assess the effectiveness of Integrated Disease Management (IDM) modules.
- Issuing of Sanitary and Phytosanitary (SPS) certificate vegetable produce for safe export/transboundary movement under trade.
- Quantify spatial and temporal pathogen populations in a specific location.
- Quantify pathogen populations in relation with regional and seasonal yield losses.

6. Postharvest fungal diseases

Common postharvest diseases resulting from wound infections initiated during and after harvest includes blue and green mold (*Penicillium* spp.) and transit rot (*Rhizopus stolonifer*). Important fungal genera of anamorphic postharvest pathogens include *Penicillium*, *Aspergillus*, *Geotrichum*, *Botrytis*, *Fusarium*, *Alternaria*,

Colletotrichum, *Phomopsis*, *Rhizoctonia*, *Sclerotium*, and *Sclerotinia*. The most important pathosystem of postharvest vegetables are gray mold (*Botrytis* spp.), white mold and watery soft rot (*Sclerotinia* spp.), cottony leak (*Pythium* spp.) and *Sclerotium* rot (*Sclerotium rolfsii*) [6].

6.1 *Sclerotinia*: rot

White mold (*Sclerotinia sclerotiorum*) appears in warm and moist weather (>95% relative humidity) and favors fungal growth on infected pods which develops as a white, fluffy mycelial mat often with large, irregular, black-colored sclerotia, typical of *S. sclerotiorum* [11–13]. Within the superficial mycelium, initially white but later hard dark black sclerotia are formed. Infected pods show brown discoloration and soft rot. The isolated fungus was identified as *S. sclerotiorum* based on morphological and cultural characteristics of the mycelia and sclerotia (**Figure 1**) [14].

6.2 *Ascochyta*: blight

Ascochyta blight (*Ascochyta pisi*) black spot symptoms on pods result in the production of round tan-colored sunken spots bearing dark margins. Pycnidia develop in the centers of such spots on pods (**Table 1**).

6.3 *Phytophthora*: late blight

Tomato (*Solanum lycopersicum,* solanaceae) is one of the most important vegetable crops. In the last couple of the years, the disease has become one of the most devastating threats to the cultivation of tomatoes in eastern Uttar Pradesh [15, 16]. Initial disease symptoms appeared in the form of irregular; water-soaked and light brown lesions on leaves which are normally covered with white cottony mycelial growth on the lower side of leaves. Water-soaked brown lesions expanded rapidly on stem and green fruits. Infected green fruits of tomato usually developed olivaceous, brown-colored leathery, and hard structures. All infected fruits eventually fall of from the plants and they were neither fit for marketing nor human consumption. Microscopic studies of the colonized pathogen on potato slices revealed hyaline, coenocytic, branched hyphae, and aseptate sporangiophores with lemon-shaped, papillate sporangia. Sporangia dimensions were $32 \pm 6.3 \times 20 \pm 4.9$ μm, with a length to width ratio of 1.6. On the basis of morphological characteristics and sporangia size, the pathogen was confirmed as *P. infestans* (**Figure 2**) [17].

6.4 *Colletotrichum*: fruit rot

Chili (*Capsicum annum*, solanaceae) is an economically important spice crop, widely grown in India. *Colletotrichum* sp. is an anamorphic fungal genera ranked in 8th position among top 10 fungal plant pathogens in the world. Infected fruits showing typical anthracnose symptoms of sunken necrotic lesions with a black dot like acervuli in concentric rings collected and collected fruit samples were examined under a light transmission microscope. Anthracnose (*Colletotrichum lindemuthianum*, *C. orbiculare*) symptoms appear on immature pods. Sunken cankers with lighter or gray central areas of about 5–7 mm size are seen. The spots on vegetable pods are enlarged and produce tiny black acervuli in the centers which in humid conditions ooze viscous droplets consisting of a mass of pinkish spores. Pure culture

Figure 1.
Typical symptoms of Sclerotinia white rot and culture plate. (A) Indian bean, (B) Indian bean, (C) French bean, (D) pea, (E) pea, (F) brinjal, (G) tomato, (H) bottle gourd, (I) PDA culture plate.

Pathogen	Disease	Symptom
Sclerotinia sclerotiorum	Watery soft rot or white stem rot	Disease symptom initially appears in the form of water-soaked lesions on pods and stems. Later, infected tissues become whitish and covered with white mycelia mats and black-colored sclerotia.
Colletotrichum lindemuthianum	Anthracnose	Disease symptoms appear in the form of brown to black sunken spots and lesions on leaves, stems, and pods. The center of anthracnose lesions on pods is covered with numerous black dot-like acervuli.
Ascochyta pisi	*Ascochyta* blight	Black spot symptoms on pods result in the production of round tan-colored sunken spots bearing dark margins with pycnidia on pods.
Macrophomina phaseolina (*Rhizoctonia solani*)	Charcoal rot or ashy stem blight	Disease symptoms appear in the form of dark brown to black charcoal-colored lesions covered with black dot-like fruiting bodies (resting microsclerotia and pycnidia) on pods.
Sclerotiorum rolfsii	Sclerotiorum rot	Whitish growth with mustard-like sclerotia on pods.
Pythium spp.	Cottony leak	White mycelial growth on pods.

Table 1.
Postharvest diseases/pathosystem of leguminous vegetable crops.

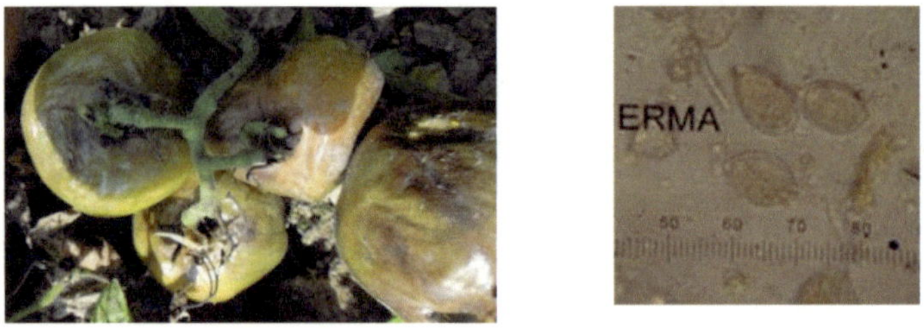

(A) (B)

Figure 2.
Typical Symptom of Phytophthora *blight on tomato fruits and Sporangia. (A) Tomato, (B) sporangia.*

(A) (B) (c) (D)

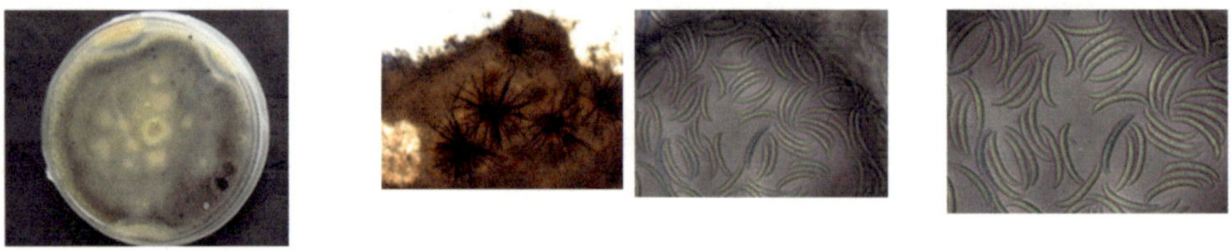

PDA culture plate Acervuli Conidia (10 x) Conidia (40 x)

Figure 3.
Typical symptoms of Colletotrichum *fruit rot (anthracnose), culture, and conidia. (A) Chili, (B) cowpea, (C) bottle gourd, (D) bottle gourd.*

of the pathogen isolate was established on PDA by the hyphal tip method. Under the light microscope, one-celled, smooth-walled hyaline falcate, tapered ended conidia (16–26 × 3–4 µm) and acervuli with numerous setae (15–27 × 2–5 µm), were recorded. In this respect, this documentation will play an important role for better understanding of the pathogen and formulation of disease management strategies for the prevention of pre and postharvest crop losses under changing climatic scenarios (**Figure 3**).

6.5 *Didymella*: blight/rot

Gummy stem blight (GSB) is caused by *Stagonosporopsis cucurbitacearum* (syn. *Didymella bryoniae*). *S. cucurbitacearum* is an airborne, seed-borne and soil-borne facultative necrotrophic plant pathogen. A black dot like pycnidia is observed on the infected fruits. Its incidence was recorded on cucurbits such as cucumber, bottle gourd, ash gourd, watermelon, etc. in the field and polyhouse. Inoculated PDA plates were produced white mycelium after 4 weeks of incubation at 24°C. Conidia were cylindrical non-septate to monoseptate and

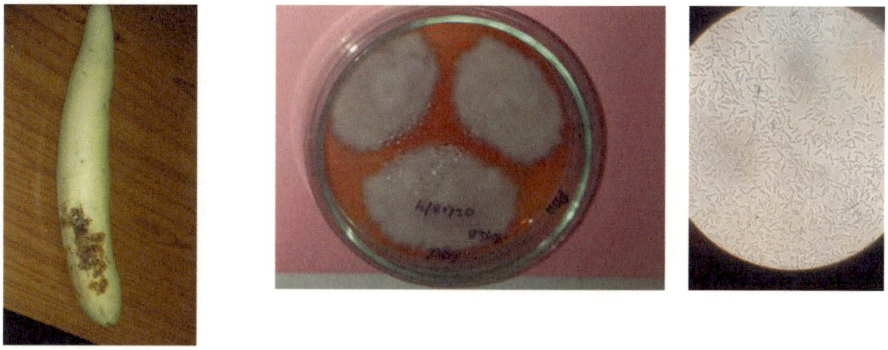

Figure 4.
Didymella black rot, culture plate, and conidia. (A) Bottle gourd, (B) PDA culture plate, (C) conidia.

Disease	Pathogen	Incidence (%)
Black rot	*Didymella bryoniae*	50
Fruit spot	*Colletotrichum laginarium*	18–23
Sclerotinia rot	*Sclerotinia sclerotiorum*	10
Blossom blight	*Choanephora infundibulifera*	30

Table 2.
Postharvest diseases/pathosystems of cucurbitaceous vegetable crops.

Disease	Pathogen	Crop	Incidence (%)
Phomopsis fruit blight	*Phomopsis vexans*	Brinjal	40–60
Sclerotinia fruit blight	*Scletrotinia sclerotiorum*	Brinjal	5–10
Fruit blight	*Phytophthora infestans*	Tomato	15
Sclerotinia rot	*Sclerotinia sclerotiotrum*	Tomato	30
Rhizoctonia rot	*Rhizoctonia solanaii*	Tomato	30
Alternaria rot	*Alternaria solani*	Tomato	30
Colletotrichum fruit rot	*Colletotrichum dematium*	Chili	20

Table 3.
Postharvest diseases/pathosystems of solanaceous vegetable crops.

60 × 40 μm in size. Based on the morphological and microscopic characteristics, the pathogen was identified as *Stagonosporopsis cucurbitacearum* (syn. *Didymella bryoniae*) (**Figure 4** and **Table 2**).

6.6 *Phomopsis*: blight

Brinjal (*Solanum melongena*, Solanacae) is one of the most important vegetables worldwide. *Phomopsis* vexans is one of the notorious seed-borne fungal pathogens that causes destructive *Phomopsis* blight which ranked second topmost disease of brinjal in India (**Table 3**). Brinjal fruit rot due to the incidence of this disease has been estimated up to 60%. The pathogen was identified on the basis of colony morphology and size of conidia (20–40 × 40 μ) (**Figures 5** and **6**).

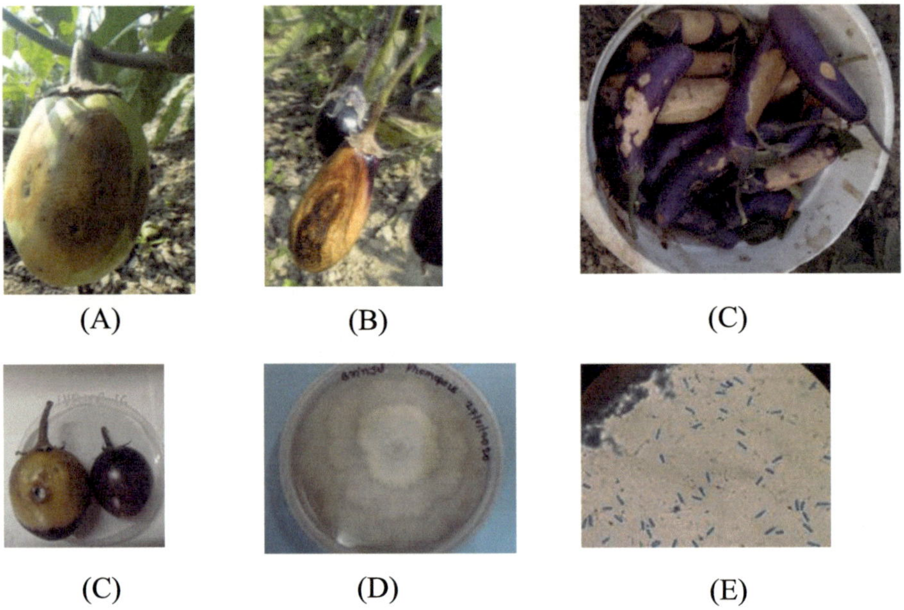

Figure 5.
Typical symptom of Phomopsis, *culture plate, and conidia. (A–B) brinjal, (Brinjal), (C) brinjal, (D) PDA cultutre plate, (E) conidia.*

Figure 6.
Alternaria fruit rot (A–C) and Sclerotinia rot, cultutre and sclerotia. (A) Tomato, (B) tomato, (C) cauliflower, (D) tomato, (E) PDA culture plate, (F) sclerotia.

7. Postharvest bacterial diseases

Phytopathogenic bacteria cause postharvest diseases of economically important vegetables. Different species of bacteria belonging to top ten genera viz. *Pseudomonas*; *Ralstonia*; *Agrobacterium*; *Xanthomonas*; *Xanthomonas*; *Xanthomonas*; *Erwinia*; *Xylella*; *Dickeya* (*dadantii* and *solani*); *Pectobacterium* are devastating plant pathogens [18, 19]. They are unable to penetrate directly into plant tissue; however, they enter through wounds or natural plant openings. Wounds can be caused by insects and tools during operations like pruning and picking of the produce. The bacteria only become more active and cause infection when factors are conducive. Factors conducive to infection are high humidity, crowding, poor air circulation, plant stress caused by overwatering, under watering, or irregular watering, poor soil health, and deficient or excess nutrients. The bacteria multiply quickly when free moisture and moderate temperatures are available. The major causal agents of bacterial soft rots are various species of *Erwinia*, *Pseudomonas*, *Bacillus*, *Lactobacillus*, and *Xanthomonas. Psuedomonas syringae* pv. *syringae*, *P. syringae* pv. *pisi* and *P. syringae* pv. *phaseolicola* causes diseases in vegetables [20]. The symptoms appear as water-soaked spots on pods that become sunken and dark-brown in color with distinctive reddish-brown margins.

Biological (culture media, diagnostic hosts, bacteriophages (phage typing); biochemical (based on properties of the bacteria in culture (gram stain, bacterial cell size, flagella), metabolic fingerprinting (API/BIOLOG system), thin layer chromatography, gel electrophoresis, conductance assays, isozyme analysis); immunoassays (agglutination, gel diffusion, ELISA, dot blot assays, immunofluorescence, flow cytometry); nucleic acid (hybridization, RFLPs, PCR, ICAN, DNA arrays, multilocus sequence typing) were used for reliable and accurate detection of plant pathogens for their effective management.

7.1 *Xanthomonas*: blight

X. campestris pv. *vesicatoria* now reclassified as *X. euvesicatoria*, the causal agent of bacterial spot of tomato. The disease is prevalent in warm, humid, and temperate regions of the world. The genus *Xanthomonas* comprises 20 different species [21] with many pathovars causing economically important diseases on different vegetable crops [22]. *Xanthomonas* is a rod-shaped, gram-negative bacterium. It produces a yellow soluble pigment, called xanthomonadin, and extracellular polysaccharide (EPS). EPS is an important pathogenicity factor of bacteria that protect from desiccation and for the attenuation of wind- and rain-borne dispersal (**Figure 7**).

7.2 *Pseudomonas*: blight

The disease is caused by pathogen, *Pseudomonas syringae* pv. *pisi*. Another bacterium, *P. syringae* pv. *syringae* has also been reported to infect vegetable pea in a temperate region. Tender pods are chocolate brown, thin, twisted, and

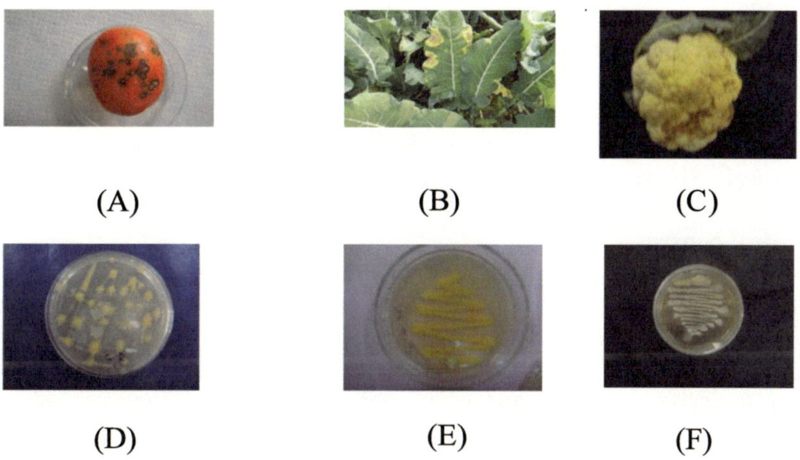

Figure 7.
Xanthomonas *blight (A and B) and* Pectobacterium *soft rot (C) and culture plate of* Xanthomonas *(D and E) and* Pectobacterium *sp. (A)* Xanthomonas *tomato speck, (B)* Xanthomonas *blight on kale, (C)* Pectobacterium *soft rot on cauliflower, (D)* Xanthomonas *NA culture plate, (E)* Xanthomonas *NA culture plate, (F)* Pectobacterium *NA culture plate.*

Crop	Disease	Pathogen	Incidence (%)
Tomato	Soft rot	*Pectinovora (Erwinia) carotovora* pv. *carotovora*	5
	Bacterial speck	*Xanthomona* sp.	5
Chili	Soft rot	*Pectinovora (Erwinia) carotovora* pv. *carotovora*	2
Beans	Soft rot	*Pectinovora (Erwinia)*, *Pseudomonas*, *Bacillus*, *Lactobacillus* and *Xanthomona* sp.	5
Cabbage	Black rot	*X. campestris* pv. *campestris*	10
Cauliflower	Soft rot	*Pectinovora (Erwinia) carotovora* pv. *carotovora*	19
Summer squash	Soft rot	*Pectinovora (Erwinia) carotovora* pv. *carotovora*	5–10

Table 4.
Postharvest bacterial diseases/pathosystem of vegetable crops.

shriveled. Lesions are large on older pods and they become thin, twisted, and dry. Seeds become discolored and shriveled. Dried bacterial ooze makes the pod surface glossy.

7.3 *Pectobacterium*: soft rot

Pectobacterium carotovorum and *Pectobacterium atrosepticum* (formerly *Erwinia carotovora* subspecies *carotovora* and subspecies *atroseptica)* causes huge losses of economically important fleshy vegetables worldwide. *P. atrosepticum* was the first genomically sequenced plant bacterial pathogen that is taxonomically related to animal pathogens. Genomic information is now available for *P. carotovorum* strains and other "former Erwinia" species now reclassified in the genus Dickeya. Geographically, *P. carotovorum* is widely distributed and causes soft rot diseases of several vegetable crops. However, *P. atrosepticum* is an economically important pathogen of blackleg disease of potato and restricted into the temperate region of the world (**Table 4**) [23].

8. Postharvest disease management

Postharvest losses in vegetables are found due to fungal and bacterial infection worldwide. New challenges are faced under trade liberalization and globalization, and serious efforts are needed to reduce these losses in vegetables.

8.1 Chemical control

Chemical fungicides are commonly used for the management of postharvest disease in vegetables. For postharvest pathogens which infect produce before harvest, the fungicides should be applied at field level during the crop season, and/or strategically applied as systemic fungicides. At the postharvest level, the fungicides are often applied to reduce infections already established in the surface tissues of produce or they may protect against infections occurring during storage and handling. Fungicides used during postharvest are actually fungistatic rather than fungicidal under normal usage. The fungicides are applied on the produce as dips, sprays, fumigants, treated wraps, and box liners or in waxes and coatings. Dip and spray methods are very common in postharvest treatments. The fungicides generally applied as a dip or spray method are benzimidazoles (e.g. benomyl and thiabendazole) against anthracnose, and triazoles (e.g. prochloraz and imazalil) and fumigants, such as sulfur dioxide, for the control of gray mold used for postharvest disease control [24, 25]. Dipping in hot water (at 50°C for 5–10 min, depending on the size of produce in combination with the fungicide) is also used for effective control of the disease. Sodium hypochlorite as a disinfectant is used to kill spores of pathogens present on the surface of the vegetable produce.

8.2 Biological control

International markets reject produce containing unauthorized pesticides, with pesticide residues exceeding permissible limits, and with inadequate labeling and packaging. Hence, biological control of postharvest diseases has great potential because postharvest environmental conditions like temperature and humidity can be strictly controlled to suit the needs of the biocontrol agent. Much information has been provided in relation to postharvest biocontrol and the problems faced by the development of commercial products [26, 27]. Biological control is used through microbes such as fungi, bacteria, actinomycetes, and viruses (bacteriophages) to control the postharvest disease of vegetables [1, 28–31]. The degree of disease control or disease suppression achieved with these bioagents can be comparable to that achieved with chemicals. As per estimates, the market of Indian bioagents is equivalent to 2.89% of the overall pesticide market in India with the worth of rupees 690 crores. It is expected to show an annual growth rate of about 2.3% in the coming years [32, 33]. In India, so far only 18 types of bio-pesticides have been registered under the Insecticide Act of 1968. Among agriculturally important microbes, *Trichoderma viride*, *T. harzianum*, *Pseudomonas fluorescens*, and *Bacillus subtilis* are the most potential bio-agents which as act as producers of biologically active metabolites like antibiotics and bacteriocin, elicitors and inducers of systemic resistance in plants. Biocontrol mediated pathogen inhibition is found to be more effective when the antagonist is applied prior to infection taking place. Antagonists which act against postharvest pathogens of vegetables by competitive inhibition at wound sites include the yeasts *Pichia* and *Debaryomces* species. Chitosan, for example, is not only an elicitor of host defense responses but also has direct fungicidal action against a range of postharvest pathogens. *Trichoderma* has potent antifungal activity against *Botrytis cinerea*, *S. sclerotiorum*, *Cortictum rolfsi*, and other important biotic stresses. Microbial pesticide active ingredients of *Streptomyces griseoviridis* K61 against bacterial soft rot, gray mold, *Phytophthora*; *Gliocladium catenulatum* against gray mold; *Candida oleophila* strain against postharvest diseases; *Coniothyrium minitans* against *Sclerotinia sclerotiorum*, *Sclerotinia minor*; *Trichoderma aspellerum* (formerly *T. harzianum*) against *Pythium*, *Phytophthera*, *Botrytis*, *Rhizoctonia*; *Trichoderma atroviridae* against *B. cinerea* and *B. subtilis* against *Botrytis* spp. is the most commonly used biocontrol agents for postharvest diseases.

Antagonistic yeast forms a biofilm to stick pathogen and parasitize on the hyphae of the pathogen. Bar-Shimon et al. [34] reported that biocontrol efficacy of yeast correlates with the production of lytic enzymes and their ability to tolerate high concentrations of salts. Further, molecular approaches were used to examine the role of glucanases in the biocontrol activity of the yeast *C. oleophila* and biocontrol activity was enhanced by overexpression of antimicrobial peptides. By early 2000, three postharvest biological products, Aspire™ (the USA and Israel), Bio-Save™ (the USA), and Yield Plus™ (South Africa) were available in the market. However, Aspire was initially involved to combine the product with a low concentration of postharvest fungicide [35] or salt solutions (1–2%) of calcium chloride or sodium bicarbonate and also with other additives which are commonly used in the food industry [36]. These products were also combined with physical treatments like hot air, curing, hot-water brushing, and combinations of the above with pressure infiltration of calcium for improvement of efficacy [37]. To increase bio-efficacy, the antagonists can also be combined with a sugar analogue (2-deoxy-D-glucose).

An effort has been made to develop two new products based on yeast antagonist *Candida saitoana* and a derivative of either chitosan (Biocoat) or lysozyme (Biocure). These products had been evaluated worldwide. They showed strong eradicative activity. The two commercial products based on the use of a heat-tolerant strain of *Metschnikowia fructicola* also contain other additives such as sodium bicarbonate. The additives are found highly effective to increase biocontrol efficacy to levels equivalent to those found with available postharvest fungicides. The product is marketed under the name ProYeast-ST and ProYeast-ORG in Israel by the company AgroGreen and found effective against rots incited by *Botrytis*, *Penicillium*, *Rhizopus*, and *Aspergillus*.

8.3 Plant essential oils

Botanical pesticides cause no adverse effects on non-target biota with biodegradability. It should be noted that most of the crops sprayed with botanical pesticides are quite safe for consumption after a short period after spraying. A large number of defensive of rich chemicals such as terpenoids, alkaloids, phenols, tannins, coumarins, flavonoids, etc. are present in plants which cause physiological effects on pathogens. These compounds have already been identified in the extracts/exudates of many plants. They have antimicrobial activities and are used for postharvest disease control.

The use of natural botanical products would be a supplement or an alternative to synthetic fungicide. Examples include 1,8-cineole, the major constituent of oils from rosemary (*Rosmarinus officinale*) and eucalyptus (*Eucalyptus globus*), eugenol from clove oil (*Syzygium aromaticum*), thymol from garden thyme (*Thymus vulgaris*), and menthol from various species of mint (*Mentha* species). The majority of research is progressing in this regard to develop plant oil-based pesticides. Therefore, essential oil-based formulations have great scope in the future to use as green pesticides as plant protectants in the integrated pest and disease management of value-added agriculture and horticulture crops.

Many exhaustive studies have been carried out on the utility of neem oil against various fungal pathogens. Its efficacy has been evaluated against fungal pathogens and found to be on par with the fungicide hymexazole in the control of the soil pathogens *Fusarium oxysporum*, *Fusarium ciceri*, *R. solani*, *S. rolfsii* and *S. sclerotiorum*. Researchers have reported *in-vitro* inhibition of 16 aromatic compounds against five major seed-borne fungal pathogens in the concentration range of 100–8000 ppm and the minimum inhibitory concentration (MIC) value for all the test fungi was 270–1704 ppm. Essential oils under commerce used as biopesticides have many problems, such as non-tariff barrier, scarcity of natural resources, need of quality control, and difficulties of registration. Some plant products have been commercialized. SPIC Science Foundation has developed a fungistatic product "Wanis" which has a single monoterpene as an ingredient and it is reportedly very effective in controlling more than 30 different types of phytopathogenic fungi. It is non-toxic to human beings and livestock. Recently, an antifungal agent by the name "TALENT", containing carvone as the active ingredient, derived from the essential oil of Carumcarvii, was commercialized. Mycotech Corporation product Cinnamite™, based on cinnamon oil, has been developed as a fungicide/miticide for glasshouse and horticultural crops. World-leading essential oil-based pesticide producing EcoSMART technologies developed EcoPCORR under the name Bioganic™ as insecticide and miticide for nursery crops, horticultural crops, and value-added crops under glasshouse conditions. The EcoSMART formulation is based on rosemary oil, viz. EcoTrol™ (insecticide/miticide), Sporan™ (fungicide) and eugenol oil formulation Mataran™ (weedicides) were classify as generally recognized as safe (GRAS). The search for antifungal agents of plant origin is important, which can further broaden the arsenal for disease management and can be used as alternatives or complementary to synthetic fungicides. These chemicals of biological origin are safe to use, and in a few cases can even be produced by farmers and rural communities. Thus plant essential oils are safe to the user and the environment and have a good potential as crop protectants and integrated pest management under organic farming and value-added agricultural and horticultural crops [38].

9. Postharvest handling operations of vegetable crops

Maintenance of hygiene in all stages of postharvest handling is critical to minimize the source of primary inoculum for postharvest diseases [39]. Produce should be harvested during the day instead of early morning. Field containers should be smoothed. Containers should be cleaned. Sterilized packing and grading equipment, particularly brushes and rollers, are used. Chlorinated water @ 100 ppm is commonly used for washing vegetables. This can be done with chlorine gas or with either liquid hypochlorite (pH 6.0–7.0). Containers should not be overfilled, which causes severe damage during stacking. Management of temperature is the most important factor to extend the shelf life of fresh vegetables after harvest. It begins with rapid removal of the field heat by using any of the following cooling methods: hydro-cooling, in-package ice, top icing, evaporative cooling, room cooling, forced air cooling, serpentine forced air cooling, vacuum cooling, and hydro-vacuum cooling. The relative humidity during storage should be maintained at about 85–95% for most fruits and 95–98% for vegetables. Transport vehicles should always be cleaned and sanitized before loading.

10. Conclusion

For postharvest disease management, various strategies such as postharvest handling systems, sanitation, and integration of botanicals/plant essential oil, microbial bioagents, and safe chemicals need to be integrated and develop integrated postharvest diseases management techniques under World Trade Organization (WTO) regime. Among them, it is expected that the knowledge of biocontrol will lead to new, innovative approaches to minimize postharvest decay of the product and it presents the best hope for the future of postharvest disease management of vegetable produce. Future research in this field will include a better understanding of the molecular basis of variability in the pathogen, pathogenesis, accurate and reliable diagnostic of the disease and to engineer novel and durable protection strategies against devastating postharvest diseases of vegetable crops.

Author details

Atma Nand Tripathi*, Shailesh Kumar Tiwari and Tushar Kanti Behera
Indian Institute of Vegetable Research (ICAR), Varanasi, Uttar Pradesh, India

*Address all correspondence to: antripathi_patho@rediffmail.com

References

[1] Tripathi AN, Meena BR, Pandey KK, Singh J. Microbial bioagents in agriculture: Current status and prospects. In: Rakshit A, Singh HB, Kumar Singh A, Singh US, Fraceto L, editors. New Frontiers in Stress Management for Durable Agriculture. 1st ed. Singapore: Springer Nature; 2020. pp. 490-499. 361-368

[2] Chikkasubbanna V. India (2). In: Rolle RS, editor. Postharvest Management of Fruit and Vegetables in the Asia-Pacific Region. Tokyo: Asian Productivity Organization; 2006. pp. 143-151

[3] Choudhury ML. Recent development in reducing postharvest losses in the Asia-Pacific region. In: Rolle RS, editor. Postharvest Management of Fruit and Vegetables in the Asia-Pacific Region. Tokyo: Asian Productivity Organization; 2006. pp. 15-22

[4] Tripathi AN. Detection and diagnosis of emerging postharvest pathogens (diseases) in vegetable crops. Book of Souvenir and abstracts. In: Bashyal BM, Das A, Kumar A, Kamil D, Hussain T, Geat N, Devappa V, et al, editors. International e-Conference on Postharvest Disease Management and Value Addition of Horticultural Crops. August 18-20, 2021. New Delhi, India: Division of Plant Pathology ICAR-IARI; 2021. p. 7

[5] Tripathi AN. Emerging diseases and their management in vegetable crops. In: Webinar on Applied Microbiology and Beneficial Microbes. August 26-27, 2021. Greenville, USA: Coalesce Research Group; 2021. p. 8

[6] Tripathi AN, Singh D, Pandey KK, Singh J. Postharvest diseases of leguminous vegetable crops and their management. In: Singh D, Sharma RR, Devappa V, Kamil D, editors. Post-Harvest Handling and Diseases of Horticulture Produce. 1st ed. London: CRC Press; 2021. pp. 387-396

[7] Ahsan H. India (1). In: Rolle RS, editor. Post-Harvest Management of Fruit and Vegetables in the Asia-Pacific Region. Tokyo: Asian Productivity Organization; 2006. pp. 131-142

[8] Klich MA. *Aspergillus flavus*: The major producer of aflatoxin. Molecular Plant Pathology. 2007;**8**:713-722

[9] Hedayati MT, Pasqualotto AC, Warn PA, Bowyer P, Denning DW. *Aspergillus flavus*: Human pathogen, allergen and mycotoxin producer. Microbiology. 2007;**153**:1677-1692

[10] Tripathi AN, Sharma P, Agarwal PC, Usha D, Hazarika BN, Tripathi SK, et al. Aflatoxins: Threat for agricultural trade and food safety. In: Prasad D, Ray DP, editors. Biotechnological Approaches in Crop Protection. New Delhi, India: Biotech Books; 2013. pp. 490-499

[11] Tripathi AN, De RK, Sharma HK, Karmakar PG. Emerging threat of Sclerotinia sclerotiorum causing white/cottony stem rot of mesta in India. New Disease Reports. 2015;**32**:19

[12] Tripathi AN, Sarkar SK, Sharma HK, Karmakar PG. Stem rot of roselle: A major limitation for seed production. Jaf News. 2013;**11**:14

[13] Tripathi AN, Sarkar SK, Sharma HK, Karmakar PG. Detection and characterization of roselle stem rot pathogen, *Sclerotinia sclerotiorum* (Lib.) de Bary and its sensitivity towards bioagents. In: National Symposium on Plant Pathology in Genomic Era. Chhattisgarh, India: Department of Plant Pathology, Indira Gandhi Krishi Vishwavidyalay, Raipur; 2014. pp. 8-9

[14] Bolton MD, Thomma BPHJ, Nelson BD. *Sclerotinia sclerotiorum* (Lib.)

de Bary: Biology and molecular traits of a cosmopolitan pathogen. Molecular Plant Pathology. 2006;7:1-16. DOI: 10.1111/j.1364-3703.2005.00316.x

[15] Tripathi AN, Pandey KK, Meena BR, Rai AB, Singh B. An emerging threat of *Phytophthora infestans* causing late blight of tomato in Uttar Pradesh, India. New Disease Reports. 2017;**35**:14

[16] Tripathi AN, Pandey KK, Rai AB, Sunil G. Late blight: An emerging disease of tomato in eastern Uttar Pradesh. Vegetable News Letter. 2016; **3**(1):4-5

[17] Drenth A, Sendall B. Practical guide to detection and identification of Phytophthora. CRC for Tropical Plant Protection Brisbane. Version 1.0. 2001. pp. 20-27

[18] Mansfield J, Genin S, Magori S, Citovsky V, Sriariyanum M, Ronald P, et al. Top 10 plant pathogenic bacteria in molecular plant pathology. Molecular Plant Pathology. 2012;**13**(6):614-629. DOI: 10.1111/J.1364-3703.2012.00804.X

[19] Strange NR, Scott PR. Plant disease: A threat to global food security. Annual Review of Phytopathology. 2005;**43**:83-116. DOI: 10.1146/annurev.phyto.43.113004.133839

[20] Tripathi AN. Bacterial diseases of vegetable crops and their management. In: Pandey KK, Rai AB, Singh B, editors. Recent Advances in Integrated Management of Pest and Disease in Vegetable Crops. ICAR-IIVR Training Manual No. 81. 2018. pp. 70-82

[21] Vauterin L, Rademaker J, Swings J. Synopsis on the taxonomy of the genus *Xanthomonas*. Phytopathology. 2000;7:677-682

[22] Young JM, Park DC, Shearman HM, Fargier E. A multilocus sequence analysis of the genus *Xanthomonas*. Systematic and Applied Microbiology. 2008;**5**:366-377

[23] Toth IK, Kenneth SB, Holevia MC, Birch PRJ. Soft rot erwiniae: From genes to genomes. Molecular Plant Pathology. 2003;**4**(1):17-30

[24] Ampatzidis Y, DeBellisL LA. Pathology: Robotic applications and management of plants and pant diseases. Sustainability. 2017;**9**(6):1010. DOI: 10.3390/su906 1010

[25] Waard D, Georgopoulos MA, Hollomon SG, Ishii DW, Leroux P. Chemical control of plant diseases: Problems and prospects. Annual Review of Phytopathology. 1993;**31**:403-421

[26] Droby S, Cohen L, Wiess B, Daus A, Wisniewski M. Microbial control of postharvest diseases of fruits and vegetables—Current status and future outlook. Acta Horticulturae. 2001;**553**:371-376

[27] Droby S, Wilson C, Wisniewski M, ElGhaouth A. Biologically based technology for the control of postharvest diseases of fruits and vegetables. In: Wilson C, Droby S, editors. Microbial Food Contamination. Boca Raton, FL: CRC Press; 2000. pp. 187-206

[28] Chaurasia A, Meena BR, Tripathi AN, Pande KK, Rai AB, Singh B. Actinomycetes: An unexplored microorganisms for plant growth promotion and biocontrol in vegetable crops. World Journal of Microbiology and Biotechnology. 2018;**34**(9):132

[29] Loganathan M, Rai AB, Pandey KK, Nagendran K, Tripathi AN, Singh B. PGPR *Bacillus subtilis* for multifaceted benefits in vegetables. Indian Horticulture. 2016;**61**(1):36-37

[30] Mohamed B, Benali S. The talc formulation of *Streptomyces* antagonist against *Mycosphaerella* foot rot in pea

(*Pisumsativum* L.) seedlings. Archives of Phytopathology and Plant Protection. 2010;**43**:438-445

[31] Pandey KK, Nagendran K, Tripathi AN, Manjunath M, Rai AB, Singh B. Integrated disease management in vegetable crops. Indian Horticulture. 2016;**61**(1):66-68

[32] Cheng XL, Liu CJ, Yao JW. The current status, development trend and strategy of the bio-pesticide industry in China. Hubei Agricultural Sciences. 2010;**49**:2287-2290

[33] Thakore Y. The biopesticide market for global agricultural use. Industrial Biotechnology. 2006;**2006**:194-208

[34] Bar-Shimon M, Yehuda H, Cohen L, Weiss B, Kobeshnikov A, Daus A, et al. Characterization of extracellular lytic enzymes produced by the yeast biocontrol agent *Candida oleophila*. Current Genetics. 2004;**45**:140-148

[35] Droby S, Cohen L, Daus A, Weiss B, Horev E, Chalutz E, et al. Commercial testing of Aspire: Abiocontrol preparation for the control of postharvest decay of citrus. Biological Control. 1998;**12**:97-101

[36] Droby S, Wisniewski M, El-Ghaouth A, Wilson C. Influence of food additives on the control of postharvest rots of apple and peach and efficacy of the yeast-based biocontrol product Aspire™. Postharvest Biology and Technology. 2003;**27**:127-135

[37] Droby S, Wisniewski M, El Ghaouth A, Wilson CL. Biological control of postharvest diseases of fruits and vegetables: Current advances and future challenges. Acta Horticulturae. 2003;**628**:703-713

[38] Tripathi AN, Gotyal BS, Sharma PK, Tripathi RK, Usha D, Biswas C, et al. Essential oils: As a green biopesticide for organic farming. In: Biswas SK, Pal S, editors. Organic Farming and Management of Biotic Stresses. New Delhi, India: Biotech Books; 2014. pp. 548-554

[39] Toth IK, van der Wolf JM, Saddler G, Lojkowska E, Helias V, Pirhonen M, et al. Dickeya species: An emerging problem for potato production in Europe. Plant Pathology. 2011; **60**:385-399

Chapter 6

Advances in Postharvest Disinfestation of Fruits and Vegetables Using Hot Water Treatment Technology-Updates from Africa

Shepard Ndlela, Nelson L. Mwando and Samira A. Mohamed

Abstract

Hot Water Treatment (HWT) provides adequate phytosanitary assurance that treated fruits and vegetables exported abroad are free from devastating quarantine pests. Two systems for HWT are currently available for commercial use namely the batch/jacuzzi and the continuous flow system depending on user requirements. Several protocols have been developed the world over and a few in Africa, but adoption has been lagging because of various factors chief among them lack of large scale validations of experiments to guide application at the commercial level. Mango, Bell pepper, avocado, and French beans play an important role in the livelihoods of people in Africa. However, their export is constrained by pests such as the invasive Oriental fruit fly, the false codling moth, and thrips. To circumvent this issue, disinfestation HWT protocols have been developed which seek to provide quarantine assurance to lucrative export markets. Hot Water Treatment technology has several advantages over other conventional phytosanitary treatments. It provides a triple function of cleaning, disinfesting, and disinfecting and is friendly to users, consumers of the treated commodities, and the environment. We discuss HWT in the context of its future and applicability in Africa. It is the future of postharvest treatments.

Keywords: Export, Mango, Preharvest, physiologically mature, Sap, Lucrative export markets, Interceptions, Quarantine, Phytosanitary, Biochemical, Physical parameters

1. Introduction

Fruits and vegetables contain important nutrients which make them vital components of a balanced diet essential for healthy living [1]. Several studies have been conducted worldwide, on the diseases and deaths occurring as a result of lack of vegetable and fruit consumption [2, 3]. Health and nutrition constitute an impending global catastrophe if clear action steps are not taken to contain the threat. The FAO reports that a staggering 38% of the world's population cannot afford the cheapest healthy meal, and for those able to at least access a healthy meal, 19% consume meals with inadequate nutrients [4].

Fruits are important sources of nutrients, minerals, and antioxidants necessary for energy, body repair, and growth as well as regulation of physiological processes [5]. The demand for fruits and vegetables is increasing globally, especially in the developed countries due to the increased push for healthy diets than just quantities [6]. The developed world has a huge deficit of fruit and vegetable supply and must supplement this with imports from the developing world such as Asia and Africa. It is estimated that the developing world produces 98% of the world's fruit production and the high-income countries in the west import over 80% of this quantity [4]. On a positive note, fruit production is rising steadily in Africa, with millions of growers at the smallholder level contributing significantly to national yield [7, 8]. Africa is endowed with a variety of indigenous fruits and imported exotic varieties cultivated for both the local and export markets [9]. However, the export of these fruits and vegetables is constrained by stringent export requirements mainly by the USA and European Union (EU). These requirements safeguard the possible entry and spread of harmful pests into areas where they are absent and will cause devastating effects once established. Upon being received at ports of entry, most fruits and vegetables are expected to be accompanied by a relevant certification from the exporting country and are then subjected to phytosanitary inspections. If the exported commodity is found to be infested by insect pests considered to be quarantine or regulated in nature, the consignment is rejected and destroyed at the exporter's cost.

The EU has major regulations that govern imports against quarantine and regulated pests such as the Regulation (EU) 2016/2031 and the Commission Implementing Regulation (EU) 2019/2072 [10, 11] among various other amendments and directives. The USA is a major player in the import of fruits and vegetables and has several federal orders and directives issued under the Plant Protection Act –7U.S.C. 7701 et seq- [12] to protect against the entry of foreign injurious pests. In addition, all countries through their National Plant Protection Organizations (NPPOs) are bound by various international standards for phytosanitary measures to abide by fair practices when exporting commodities e.g. ISPM 20 offers guidelines for an import regulatory system [13].

Though effective pre-harvest management techniques for pests of fruits and vegetables are widely available, they are unable to confer 100% freedom from infestation. In this regard, postharvest treatment technologies can provide quarantine assurance and allow exporters to meet export requirements and satisfy the growing international demand for fresh fruits and vegetables [14]. Various postharvest treatment technologies are available for adoption in providing quarantine assurance that fruits and vegetables are free from devastating pests not found in importing countries. These include controlled atmospheres [15], irradiation [16] Vapor Heat Treatment (VHT) [17], cold treatment [18], and immersion Hot Water Treatment (HWT) [19, 20] among others. Though the treatments may achieve the objective of disinfecting and disinfesting fruits and vegetables, they are also known to affect the physical and chemical properties of treated commodities [21]. This chapter explores HWT with a specific focus on Africa and applications aimed at satisfying requirements by lucrative export markets such as the USA and the EU. Prospects for commercial application of HWT on mango, avocado, bell pepper, and French beans are discussed.

2. Constraints to export necessitating HWT technology

Data on the loss of export earnings in most African countries is scarce due to poor coordination, lack of financial resources to conduct systematic studies, and lack of standard indicators for measurement of loss in mixed production cropping systems and value chains. Currently, most African countries are not exporting or are exporting under fear of interceptions. Following the invasion of Africa by the devastating polyphagous fruit fly *Bactrocera dorsalis* (Hendel) (Diptera: Tephritidae), [22] the USA through an emergence Federal order restricted the movement into the USA of fruits and vegetables deemed to be hosts of the pest [23]. The fruits and vegetables were only to be granted entry if they were originating from a country free from *B. dorsalis*, grown under a structure that excludes pests as certified by the USA systems approach 7CFR319.56, or having been subjected to an approved treatment [23]. The list of hosts was updated in 2015 through another Federal order DA-2015-2054, making exports nearly impossible for African countries where the pest continues to spread rapidly. Recently the EU announced new regulations on mango imports, through directive (EU) 2019/523 of 21 March 2019 which largely affects developing countries particularly from Africa where both pre-and post-harvest systems are absent or poorly developed. The directive stipulates that for mangoes to be accepted in the EU, the consignment must originate from a country free from quarantine fruit flies, a pest-free area, or a pest-free production site or show empirical data that is effective postharvest treatment has been administered [24].

To this effect, several countries have suffered losses in potential export earnings due to the unavailability of postharvest treatments that would allow them to export mango, either regionally or internationally. For example, as a way of protecting the vast fruit production industry in South Africa, from the *B. dorsalis* menace at the height of the incursion, the government halted the importation of fruits from Mozambique resulting in the former Portuguese colony suffering a USD 2.5 million revenue loss [25]. A single export company in Mozambique reported losses of USD 1.5 million per year due to restrictions on exports of fruits and vegetables from Mozambique [26]. Kenya similarly lost USD 1.9 million worth of export earnings due to an import ban imposed by South Africa [27]. In West Africa, potential export losses due to *B dorsalis* are estimated at USD 220 billion per annum [28]. This is huge considering that millions of families derive their livelihoods from the various value chains in the fruit and vegetable production sector.

The recent insurgency by the native invasive False codling moth (FCM), *Thaumatotibia leucotreta* (Meyrick) (Lepidoptera: Tortricidae) has also resulted in thinning of export markets due to quarantine restrictions [29]. The pest threatens the production of avocadoes, capsicum, citrus, and various solanaceous crops. The FCM is highly polyphagous, and is native to Africa but is known to occur in Israel [30, 31]. Interception of commodities at ports of entry is on the rise with pests having been intercepted several times in the USA but with no record of establishment [32]. There are also fears that it could invade Europe, thus restrictions have been heightened by the EU [33]. Through the European Commission Implementing Directive 2017/1279, the new regulations require that capsicum be exported from pest-free areas or be subjected to an effective treatment before export. A systems approach of managing these devastating pests is not sustainable in fragmented production systems as has been mentioned afore, and similarly, growing these crops under protected environments is expensive and beyond the reach of resource-poor smallholder growers in Africa, leaving the option of postharvest treatment as the most feasible and efficient approach because it can be done by medium to large scale commercial enterprises.

2.1 Hot water treatment technology

Hot Water Treatment technology particularly immersion, involves subjecting fruits or vegetables to hot water at a specified temperature and duration beyond the thermal limits tolerated by the various stages of development for pests of interest. The temperature and duration used are often that which kills the most thermotolerant stage of development. The preference for HWT emanates from the fact that it is an efficient system in heat transfer to treated commodities than for example air due to its capacity to conduct heat [34]. It may not need specialized structures except for temperature regulation, water circulation, and a simple vessel to hold water during treatment which makes it ideal even for small-scale applications [21]. Furthermore, HWT performs a dual function of disinfecting and disinfecting in one treatment.

Immersion HWT is not new in phytosanitary treatment spheres. Historically it was mainly used in killing pathogens on fruits but has since found new applications in killing insect eggs and larvae on or in fruits [35]. Protocols to disinfect and disinfest different commodities of economic value have been developed in various countries worldwide [28, 36]. Some of the protocols developed for mango, bell pepper, and avocado at the international level (**Table 1**) have been instrumental in guiding similar research in Africa where the three commodities have the potential to boost foreign earnings significantly if exported to lucrative markets. The HWT technique has become an integral alternative for postharvest treatments due to the demand for non-chemical treatments against pestiferous insects and pathogens [35]. Consumers are increasingly becoming aware of their health and progressively seek alternatives for treatments whose effects are presumably unknown at present, or are shrouded in mystery and conspiracy theories. The technique is equally important as a forerunner of other treatments for example HWT has been shown to decrease the effects of cold treatment on fruit and vegetables [54]. Commodities which are mostly exported in refrigerated containers overseas for long periods are sometimes susceptible to cold injury, hence short periods of immersion in HWT induce some level of tolerance to cold and increases quality and shelf life [59]. In bell pepper, heat treatment increases the levels of polyamines which are responsible for reducing cold injury and decay, thus allowing the capsicum to be stored for longer periods than would be possible when subjected to cold treatment alone [54].

The ecotoxicological effects [both known and unknown] of various chemicals used in postharvest treatments or their excessive use at preharvest meant to produce clean fruits without the need for postharvest treatments have also caused a renewed interest in HWT. Some chemicals from treatment effluent or evaporation are known to stress ecological systems leading to irreversible damage [60]. At the body level, exposure and effect regarding the endocrine, reproductive and other systems are known to be current concerns and emerging issues [61]. Hot Water Treatment technology offers a non-chemical system of treating fruits and vegetables which is friendly to the users, consumers, and the environment.

Treatments aimed at disinfecting, by cleaning with pure water, fungicide, bleaching liquids, or some generally recognized as safe (GRAS) compounds are often for short periods [62, 63]. Those aimed at disinfecting or assuring freedom from quarantine or regulated insect pests take long periods usually up to one hour or more [35, 64].

2.2 Requirements for HWT

In most countries where HWT is practiced, the design of the treatment machine and the entire facility is chosen by the individual or company wishing to treat fruits and vegetables. The current technology uses the batch/jacuzzi system, the continuous flow system (CFS) [65], and the drainage system [66]. The continuous flow system is mostly used in the disinfecting system due to the short periods that fruits have in contact with water, while the jacuzzi system allows fruits to be submerged in hot water for long periods making it ideal for disinfecting purposes. In the batch

Commodity	Treatment protocol	Target	Reference
Mango	46.1–46.7 °C for 45–65 min	*Anastrepha suspensa*[1]	[37]
	46.1–46.7°C for 54 min	*A. suspensa*	[38]
	46.1°C for 90 min	*Anastrepha ludens*[2], *Anastrepha obliqua*[3]	[39]
	46.1–46.7°C for 60–65 min	*A. obliqua, A. suspensa*	[40]
	46.1–46.7°C for 90 min.	*A. suspensa*	[40]
	46°C for 60 min; 46°C for 90 min; 49°C for 120 min	*A. suspensa*	[41]
	48°C for 7.5–30 min	Quality test	[34]
	48°C for up to 90 minutes	*Bactrocera [Dacus] aquilonis*[4]	[42]
	47°C for up to 2 hr. [core temperature]	Quality control	[43]
	46.1–46.7°C for 20–60 min.	*A. obliqua*	[44]
	50°C for 30 min	Quality control	[45]
	45°C for 75 min	Quality control	[46]
	44°C for 25 min [core temperature]	*A. suspensa*	[47]
	48°C for 60–75 min.	*Bactrocera dorsalis*[5]	[48]
	46.5°C for 90 min through phased hot water	Quality control	[49]
Avocado	41°C for 25–30 min, or 42°C for 25 min after cold treatment	Fruit flies	[50]
	38 °C for 1 hr.; 50°C for 1-10 min [pretreatment]	Quality control	[51]
	38 °C for 60 min [pretreatment]	Quality control	[52]
	38 °C for up to 120 min pretreatment for 50°C for 10 min	Quality control	[52]
Bell pepper	45°C for 15 min, 53°C for 4 min	Fungus	[53]
	53°C for 4 min	Quality control	[54]
	55± 1°C for 12 ± 2 s.	Fungus	[55]
	40, 50, 60°C for 2 mins	Fungus	[56]
	53°C, 1 to 3 mins	Quality control	[57]
	55°C for 15 sec	Quality control	[58]

1–5 Are fruit flies [Diptera: Tephritidae].[1]*The Caribbean fruit fly: A quarantine pest of various fruits including mango, citrus, guava, Annona.*
[2]*The Mexican fruit fly: invasive pest, polyphagous attacking mango and citrus among other host plants.*
[3]*West Indian fruit fly: Invasive pest attacking a wide range of fruits among them mango.*
[4]*Very destructive fruit fly with a very broad host range.*
[5]*Oriental fruit fly-devastating invasive fruit fly first reported in Africa in 2003, attacks a wide range of fruits and vegetables.*

Table 1.
Some immersion hot water treatment protocols for disinfesting and disinfecting mango, bell pepper, and avocado developed worldwide.

system, fruits are lowered into large tanks containing circulating hot water either manually or by hydraulics. They have only removed the same way when treatment duration has been completed. On the other hand, the drainage system uses hot water which wets fruits from above as they move on a stainless steel conveyor belt [66]. This is also ideal for disinfecting fruits and vegetables. Due to the large volumes of treated fruit required at commercial scale [in both treatments for disinfecting or disinfesting], the jacuzzi system is slow hence the preference for the CFS which is automated. An ideal HWT treatment equipment must be equipped with a heating mechanism that supplies continuous uniform heat, temperature sensors for temperature regulation, uniform circulation of heated water.

Hot water treatment of fruits is largely dependent on the variety, shape, size and, physiological maturity of fruits. This makes it a garbage in, garbage out (GIGO) process. We strongly opine that most quality challenges in treated commodities emanate from poor selection and handling during harvesting, treatment, and post-treatment. The desire to have fruits last long periods during transit often forces fruit growers to harvest physiologically immature mangoes, which however fail to tolerate heat leading to affected esthetics and physiological internal quality.

2.3 Protocols for HWT developed in Africa

Various protocols particularly for disinfecting and disinfesting mangoes, bell pepper, and avocadoes have been developed in Africa (**Table 2**) but their applicability in real commercial entities is restricted due to various reasons ranging from effects on quality, ease of application, set up cost, and commitment from relevant sectors. Between 2014 and 2019, approximately 29 million tonnes of mango were produced by the top five counties in Africa (**Table 3**), and most of this produce was consumed locally or found its way to less stringent markets such as the Middle East, which are sadly less lucrative.

No commercial HWT facilities are established except for miniature systems used for experimentation in the top twenty producers of mango making it difficult to export to the USA or EU without risking costly interceptions. There are reports

Commodity	Treatment protocol	Target	Reference
Mango	46.1 °C for 81.47 min [95% CL 75.77–87.18 min], with 68 mins equally effective	*Bactrocera dorsalis*	[19]
	46.1°C for 86.7 min [95% CL 77.830–99.880 min] with	*B. dorsalis*	[20]
	52°C for 5 minutes	anthracnose	[67]
	Fruit core temperature of 46.5°C	*B. dorsalis*	[68]
Avocado	38°C for 15–30 min	pathogens and external chilling injury	[69]
	46°C for 5 min	Quality control	[70]
	40°-42°C for 20–30 minutes.	Quality control	[71]
	38 °C for 30 minutes, 42°C for 25 minutes and 46°C for 20 minutes	Quality control	[72]
	38 °C for 5 minutes,	Quality control	[73]
Bell pepper	50°C for 3 min plus Controlled atmosphere packaging	Quality control	[74]
French bean	50 °C for at least 5 min	*Frankliniella occidentalis**	[75]

**Frankliniella occidentalis: Commonly called the Western Flower Thrips, highly polyphagous and invasive [it is a quarantine pest]. Due to insecticide resistance, it is difficult to control through preharvest management options, hence post-harvest through HWT is very effective.*

Table 2.
Some protocols for disinfecting, disinfesting, and enhancing the quality of mango, avocado French bean, and bell pepper in Africa.

County	Mango production [metric tonnes][‡]	Country	Avocado production [metric tonnes][†]
Malawi	8,985,332	Kenya	1,347,713
Nigeria	5,522,388	South Africa	562,589
Sudan	4,962,904	Malawi	557,444
Kenya	4,777,695	Cameroon	444,066
Mali	4,400,374	DRC	360,374
Tanzania	2,215,309	Ethiopia	339,854
Madagascar	1,483,336	Côte d'Ivoire	185,628
DRC	1,242,251	Madagascar	158,828
Niger	1,065,111	Eswatini	56,979
Guinea	1,064,577	Ghana	54,665
Senegal	781,636	Congo	48,411
South Africa	544,177	Rwanda	47,847
Ethiopia	510,148	CAR	44,624
Ghana	509,639	Zimbabwe	15,532
Côte d'Ivoire	471,680	Zambia	3,941
Uganda	451,357	Réunion	3,531
Congo	182,075	Eritrea	954
Mozambique	175,803	Seychelles	65
Chad	174,792		
Sierra Leone	143,996		

[‡]*High production of mango.*
[†]*Avocado which could be exported to lucrative export markets if hot water treatment facilities were readily available. DRC: Democratic Republic of Congo; Congo: Congo Brazzaville; CAR: Central African republic.*

Table 3.
Top twenty producers of mango and avocado in Africa between 2014 and 2019 [76].

that Mozambique subjects its mangoes for export to South Africa, to HWT at 47°C for 12 minutes, [77] but it is unlikely that this treatment is adequate to disinfest any fruit fly or insect pest of mango. It has been demonstrated that at 46.1°C for durations less than 68 minutes, *B. dorsalis* larvae can still survive in apple mango variety weighing up to 500 g [19]. The most thermotolerant stage of development of this pest was only killed by continuous exposure to hot water treatment for a minimum of 68 minutes.

2.4 HWT protocols for mango

Malawi, Nigeria, Sudan, Kenya, Mali, Tanzania lead in mango production and all are struggling to satisfy requirements for the USA and EU market. Systems approach centering on preharvest management of pests often fails due to the nature of production in Africa. The systems approach works best when area-wide-IPM is practiced. Sudan does not use immersion HWT but has been exporting mangoes subjected to Vapor Heat Treatment (VHT) [78] for the past eight years now. The Sudanese Center for Sterilization of Horticultural Exports [SCS] was established in 2013 specifically to offer VHT services to the horticulture sector. The Centre is unique in that it is the first institution in Africa and the Middle East to offer such services. Mangoes are subjected to 95% Relative Humidity (RH) to achieve pulp temperature of 46–48°C for 30 minutes.

Hot Water Treatment experiments conducted in Ghana on the Keitt mango cultivar at 52°C for 5 minutes was effective in controlling anthracnose for 21 days in storage [67]. The treatment did not affect total soluble solids, titratable acidity, pH, and firmness thus was deemed adequate for adoption by the private sector. However, only 70 mangoes were used in the trial, and the equipment used was not mentioned. This has been the problem with some protocols because temperature regulation and water circulation are key yet they are not mentioned in the experimental design.

Similar experiments were also conducted in Benin on Kent cultivar at various temperature and duration regimes, but the authors recommended a treatment that results in a core temperature of 46.5°C as the most effective to provide quarantine security against *B. dorsalis* in export mango [68]. They also proposed a lower temperature between 42.0°C and 46.5°C as being similarly effective and preferable as it would result in reduced energy costs to implementers. Their treatment protocol resulted in some effects on fruit quality, and they proposed pre-conditioning fruits by subjecting them to heat before treatment and, hydro cooling the fruits after treatment to ameliorate quality challenges.

Further trials were conducted in Kenya on Apple mango at a temperature of 46.1°C for 81.47min with 95% confidence limits set at 75.77–87.18 min [19]. However, efficacy data consistently showed that 68 mins were equally effective in disinfesting mangoes of up to 500 g against the devastating *B. dorsalis*. This work also reported thermal tolerances for the various stages of development viz. eggs, first, second and third instars, with the third instars being the most heat tolerant stage of development. The protocol has since been submitted to the EU by the Kenya Plant Health Inspectorate Service (KEPHIS), the National Plant Protection Organization responsible for the regulation in Kenya, and duly recognized as an effective treatment. A pilot mango consignment treated at 46.1°C for 68 min was shipped to Italy in July 2021 by a renowned exporter Fresco Freshpro Limited and passed all phytosanitary requirements as well as quality concerns with consumers.

Recently, another protocol was developed for the disinfestation of *B. dorsalis* in Tommy Atkins mango cultivar in Uganda [20]. The temperature was set at 46.1°C and mangoes used weighed between 500 and 700 g. The ideal treatment duration was determined at 86.7min with 95% confidence limits at 77.830–99.880 min. The most heat tolerant stage was the third instar and the egg was the least tolerant just as in the study conducted on apple mango in Kenya.

These protocols provide an ideal alternative to postharvest treatments and may be adopted by countries wishing to use HWT for disinfestation and disinfection.

2.5 HWT protocols for avocado

Kenya and South Africa lead in the production of avocado (**Table 3**), however cold treatment in transit (commercial shipping temperature of approx. 5.5°C for up to 28 days) is mostly used or in some instances, the fruits are ripened, peeled, and frozen before export. Most South African Avocadoes are exported to Europe under the SADC/EU Economic Partnership Agreement and the Southern African country is making efforts to secure markets in the USA, India, Japan, and China [79]. Though cold treatment in transit offers adequate phytosanitary assurance against quarantine pests, it affects the quality of the fruits through chilling injury [71]. Several investigations have been conducted to ameliorate cold injury through the use of various treatments which include short exposures to HWT. However, there are many views regarding the efficacy and applicability of HWT in avocado [69] and trials are still being conducted to date [72].

At the peak of the *B. dorsalis* incursion in Africa, experiments were conducted to determine a cold treatment protocol to disinfest avocado from the devastating quarantine pest. Results indicated that a temperature of 1.5°C or lower, applied continuously for 18 days was adequate to kill all stages of development of *B. dorsalis* and offer phytosanitary security against the risk of introducing the pest in new territories [18]. Though these findings were in agreement with similar protocols requiring avocado to be subjected to temperatures of 1°C for 20 days [80] or 1.1–2.2°C for 14–18 days as was prescribed by USDA then, issues of fruit quality due to chilling injury remained unaddressed. Attempts to validate the 1.5°C for 18 days protocol at the Horticultural Crops Directorate (HCD) center in Kenya yielded mixed results as fruits were damaged by cold injury. Opinion has it that avocadoes must not be subjected to temperatures lower than 5°C or greater than 10°C after harvest [81]. However, this could be a major challenge considering the production system in Africa outside South Africa, where temperature-controlled systems are poorly developed. The commercial shipping temperature in South Africa is 5.5°C for 28 days, thus it could be possible that temperatures lower than this threshold may be detrimental to avocado quality.

Several studies in South Africa, have confirmed that HWT for shorter periods may confer some tolerance to cold and result in less injury than what is currently experienced and additionally disinfect the fruits against major pathogens (**Table 2**). For example, a 40°-42°C for 20–30 minutes HWT regime was found adequate to reduce chilling injury but specific protocols were required for each variety and also tested at various levels of maturity [71]. Inconsistencies in findings have largely contributed to a reluctance to adopt some protocols. For example, HWT at 46°C for 5 min produced inconsistent results in two tested varieties namely Hass and Fuerte [70]. The same treatment had given promising results in Australia, with positive physical appearance qualities after treatment and storage at 1°C for 16 days. The treatment also adequately disinfested the fruits from the Queensland fruit fly *Bactrocera tryoni* [82].

Experiments conducted at 38°C for 15–30 min yielded inconclusive results and still required further tests though the regime reduced the severity and occurrence of pathogens, and also cold injury [69]. Hot water treatment at 38°C for 5 minutes also promises to deliver the required quality but only when accompanied by waxing, special packaging using low-density polyethylene, and stepping down storage temperature from the conventional 5.5°C to 4.5°C [73]. The chilling injury was reduced and uniform ripening improved when HWT was conducted at 38°C for 30 minutes, 42°C for 25 minutes, and 46°C for 20 minutes [72]. The treatments outlined above require harmonization and validation for specific varieties at various levels of maturity. This will enable a holistic approach to the attainment of export quality through both disinfection and disinfestation. Avocadoes are susceptible to pathogens and pests like the false codling moth. They are bulky thus cannot be exported by air due to the cost connotations thus export by sea is the most feasible and cost-effective way. A sharp balance that guarantees freedom from pathogens and insects at the same time ensuring that avocadoes reach their destination without physical damage and deterioration in esthetic value must be struck if Africa is to compete at the world stage. This looks possible if correct HWT parameters are established, validated, and adopted.

2.6 HWT protocols for bell pepper

Not so many references are available in the literature for either disinfection of pathogens or disinfestation of insects in bell pepper in Africa most probably due to its susceptibility to heat and poor equipment to conduct such delicate work. At the world stage a couple of references are available notably on enhancing quality before cold treatment or disinfection [51–57]. In the African context, bell pepper was subjected to HWT at 50°C for 3 min before being stored in controlled atmosphere packaging [74]. This was adequate to kill pathogens, slow down weight loss, and loss of carotenoids and ascorbic acid.

More research is being conducted in Kenya to develop protocols for providing phytosanitary security against the devastating native invasive pest the false codling moth *T. leucotreta.* Results are promising and the private sector is eagerly waiting (Mwando et al. unpublished data a).

2.7 HWT protocols for French beans

French beans are mostly exported fresh and the western flower thrips *F. occidentalis* pose a great risk of being transported to new areas where they could potentially cause huge damage. This is so because the pest has developed resistance to major classes of insecticides thus preharvest management techniques alone are inadequate [83]. The pest poses a double not only causes direct damage but also vectors deadly viruses which are a huge threat to the horticulture industry [84]. A pilot trial to disinfest French beans using HWT was conducted at 50°C for 5 min and was found effective in causing 100% mortality of thrips eggs. The treatment did not affect the quality of French beans. Further validation has been conducted to establish a feasible protocol that can be adopted by the private sector (Mwando et al. unpublished data b).

3. Challenges affecting hot water treatment technology

Though various protocols have been developed in Africa and elsewhere, adoption by the private sector has been slow and absent in many parts of Africa. Protocols have remained on shelves unused. Most of these protocols have been developed used using water baths using very small sample sizes which can hardly be extrapolated to commercial scales. In some cases, the studies are silent on the equipment used and how heating, water circulation, and temperature regulation were achieved yet these are vital components of any HWT system. The stage of maturity of fruits is another factor that largely affects the quality of treated fruits. Most often, immature fruits are harvested by growers in a bid to ensure that they stay longer on the shelf. Maturity indices are also lacking and growers rely on visual characteristics which are not always accurate. The aspect of maturity is bound to be the sticking point because avocadoes and mangoes are bulky and are likely to be exported by sea to reduce transport costs, thus fruits are bound to be harvested earlier before reaching full physiological maturity to withstand the long voyage to Europe, America, India or China.

Resources are a limiting factor in adopting HWT in Africa with various small to medium scale enterprises unable to gather adequate startup capital. However, in our experience, we have noticed that some private sector partners have misplaced expectations, and require out-of-this-world automated facilities yet the HWT itself is simple and affordable. We opine that it is far much cheaper and easier to run and maintain than a VHT facility of similar size and capacity. Hot water treatment protocols for disinfecting fruits and vegetables usually prescribe a longer treatment duration which may be put off to potential users. The insistence of probit 9 as the required quarantine efficacy level frequently overestimates treatment times. This can be evaded by adopting other statistical analyses which equally show how effective a treatment is.

The emergence of less stringent markets in the Middle East has also provided local fruit growers and exporters with alternative markets where they send their products with less hustle. Thus investing in HWT equipment is seen as an expensive venture.

The lack of holistic protocols validated at a large scale has also been another impediment. There is a need to validate the most promising protocols and seek buy-in from major industry players if HWT is to be fully adopted at a commercial scale. Poor circulation of water and regulation of temperature also impact the quality of fruit. Hot water treatment requires a sizeable investment into good probes, data loggers, and automated controls. Too much fluctuation of temperature during treatment can be detrimental to the final product. Combination treatments are poorly developed or are simply expensive to implement. In cases where HWT cannot be a stand-alone treatment due to the susceptibility of the commodity to heat, combination treatments may then be used to circumvent the challenge. If there is a mismatch in heat tolerance between the commodity and pest, for example in instances where the pest is tolerant to higher levels of heat which cannot be tolerated by the fruit or vegetable, then HWT will not be feasible [85]. Thus the above suggestion may work.

4. Benefits of HWT

The biggest benefit of HWT is that it has a triple function of cleaning, disinfecting [86], and disinfesting [19, 20, 37]. During the treatment process, dirt, debris, and any extraneous material are removed which could potentially affect fruit or vegetable quality. Most systems incorporate a hot water brushing and rinsing station where fruits are scrubbed and rinsed dry by cool air [87]. This stage is absent for example in VHT and fungicides are easily applied if need be at this stage. The technology is relatively uncomplicated, uses clean water from conventional water sources, and is purely a non-chemical postharvest disinfestation process [88]. The cost of setting up a HWT facility is considered far much less (10%) than setting a similar VHT facility [89]. Treatment by VHT often causes scalding if vapor is not uniformly distributed, while a simple water pump can easily distribute heat uniformly in immersion HWT. Compared to irradiation and chemical treatments, HWT does not use chemicals, hence leaves no residues on treated commodities ([88] and references thein). When chemicals are used in the disinfestation of fruits e.g. the use of methyl bromide fumigation, residues are left on the fruits which pose health risks to consumers. Consumerism is increasing all over the world and consumers are becoming aware of their rights and continuously demand to be served healthy food.

Hot water treatment slows down ripening in fruits thus increasing shelf life [90]. Shelf life is very important as fruits and vegetables are perishable [74]. From harvest to consumption there is a huge lag that requires that fruits be kept in good quality to enable transportation and storage before reaching consumer tables. Experiments have also shown that HWT can confer tolerance of low temperature (chilling injury) in some fruits such as avocado [72]. Short treatments of fruits before cold treatment have been shown to reduce injury caused by excessive cold. Subjecting fruits to HWT before the actual treatment (particularly cold) is becoming more applicable because of the positive benefits.

5. Prospects

Hot water treatment may be the future of sustainable postharvest treatment of fruits and vegetables in Africa. Thus more investment into precision equipment is required. Equipment with state-of-the-art sensors, heating, and circulation apparatus. This will ensure that research is conducted in commercial facilities allowing large-scale validations to be performed. Many deformities, injuries, and effects on sensory parameters are results of poor equipment being used. Research at the molecular level is required to determine heat regimes that are not detrimental to finer qualities at the micro and macro protein levels. Combination treatments must be explored especially for heat susceptible fruits and vegetables. The combination treatments must be relatively affordable. With the wider acceptance of insects for food, the use of chitin in coating to increase shelf life must be explored [91].

It is imperative to develop specific protocols that consider variety, size, maturity, and other factors than use blanket treatments that produce horrible results. This may be costly initially but very profitable in the long run. Standards will eventually have to be developed, so that uniformity in operations is maintained. The whole system requires a Multi-stakeholder approach than lone attempts in such a global village. In such a case, customization and harmonization of protocols become feasible resulting in cost-saving especially in instances where fruit varieties/cultivars are similar and there are no significant differences in size and other physical and physiological characteristics.

6. Conclusions

Hot water treatment is the future of postharvest treatment of fruits and vegetables in Africa. Africa is a potential giant to feed the world with fruits and vegetables considering the favorable climatic conditions favoring production. However, the huge threat posed by devastating invasive pests hinders Africa from exporting to lucrative exports markets such as the USA and Europe. Several protocols have been developed for disinfestation and disinfection of commodities by their adoption far lags. Besides providing phytosanitary security, HWT can also enhance fruit quality by activating polyamines and heat shock proteins that enrich fruit and vegetable shelf life. Africa is indeed on the rise, to implement HWT technology and export large volumes of fruits and vegetables thereby increasing her foreign currency earnings.

Acknowledgements

We gratefully acknowledge the financial support for Hot water postharvest treatment work at the International Centre of Insect Physiology and Ecology -*icipe* by BioInnovate Africa grant number: BA-C1-2017-06_icipe, the Australian Centre for International Agricultural Research (ACIAR), and the International Development Research Centre (IDRC) through the project 109040; and the core funding to *icipe* by the following organizations and agencies: UK's Foreign, Commonwealth and Development Office (FCDO); the Swedish International Development Cooperation Agency (Sida); the Swiss Agency for Development and Cooperation (SDC); the Federal Democratic Republic of Ethiopia; and the Government of the Republic of Kenya. The views expressed herein do not necessarily reflect the official opinion of the donors.

Conflict of interest

The authors declare no conflict of interest.

Author details

Shepard Ndlela*, Nelson L. Mwando and Samira A. Mohamed
International Centre of Insect Physiology and Ecology, Nairobi, Kenya

*Address all correspondence to: sndlela@icipe.org; shepardndlela@gmail.com

References

[1] Pomerleau J, Joint F, World Health Organization. Effectiveness of interventions and programmes promoting fruit and vegetable intake [Internet]. researchonline.lshtm.ac.uk. [cited 2021 Aug 19]. Available from: https://researchonline.lshtm.ac.uk/id/eprint/13792/

[2] Boeing H, Bechthold A, Bub A, Ellinger S, Haller D, Kroke A, et al. Critical review: Vegetables and fruit in the prevention of chronic diseases. Eur J Nutr. 2012;51(6):637-663.

[3] Mo X, Gai RT, Sawada K, Takahashi Y, Cox SE, Nakayama T, et al. Coronary heart disease and stroke disease burden attributable to fruit and vegetable intake in Japan: Projected DALYS to 2060. BMC Public Health. 2019;19(1):1-9.

[4] FAO, IFAD, UNICEF W and W. Review of The State of Food Security and Nutrition in the World, 2019. Vol. 10, World Nutrition. 2020. 95-97 p.

[5] Maldonado-Celis ME, Yahia EM, Bedoya R, Landázuri P, Loango N, Aguillón J, et al. Chemical Composition of Mango (Mangifera indica L.) Fruit: Nutritional and Phytochemical Compounds. Front Plant Sci. 2019;10(October):1-21.

[6] Mason-D'Croz D, Bogard JR, Sulser TB, Cenacchi N, Dunston S, Herrero M, et al. Gaps between fruit and vegetable production, demand, and recommended consumption at global and national levels: an integrated modelling study. Lancet Planet Heal [Internet]. 2019;3(7):e318–e329. Available from: http://dx.doi.org/10.1016/S2542-5196(19)30095-6

[7] Temu AE TA. High Value Agricultural Products for Smallholder Markets in Sub-Saharan Africa: Trends, Opportunities and Research Priorities Prepared for. In: International Workshop on how can the poor benefit from the growing markets for high-value agricultural products? Cali, Colombia.: International Center for Tropical Agriculture; 2005. p. 1-37.

[8] Joosten F, Dijkxhoorn Y, Sertse Y, Ruben R. How does the fruit and vegetable sector contribute to food and nutrition security? [Internet]. Wageningen; 2015 [cited 2021 Aug 20]. Available from: https://library.wur.nl/WebQuery/wurpubs/489413

[9] Jamnadass RH, Dawson IK, Franzel S, Leakey RRB, Mithfer D, Akinnifesi FK, et al. Improving livelihoods and nutrition in sub-Saharan Africa through the promotion of indigenous and exotic fruit production in smallholders' agroforestry systems: A review. Int For Rev. 2011;13(3):338-354.

[10] EU. Commission Implementing Regulation (EU) 2019/2072 of 28 November 2019 establishing uniform conditions for the implementation of Regulation (EU) 2016/2031 of the European Parliament and the Council, as regards protective measures against pests of plants, and repealing Commission Regulation (EC) No 690/2008 and amending Commission Implementing Regulation (EU) 2018/2019. Off J Eur Union [Internet]. 2019 Dec 10 [cited 2021 Aug 20];L 319/1:1-279. Available from: https://eur-lex.europa.eu/eli/reg_impl/2019/2072/oj

[11] Black R, Bartlett DMF. Biosecurity frameworks for cross-border movement of invasive alien species. Environ Sci Policy [Internet]. 2020;105(October 2019):113-9. Available from: https://doi.org/10.1016/j.envsci.2019.12.011

[12] USDA. Plant Protection Act (As Amended Through Public Law 108-498, Dec. 23, 2004).

USDA-APHIS-PPQ-Professional Dev Cent [Internet]. 2004;(June). Available from: https://www.aphis.usda.gov/plant_health/downloads/plant-protect-act.pdf

[13] FAO-IPPC. Adopted Standards (ISPMs) - International Plant Protection Convention [Internet]. Available from. 2021 [cited 2021 Aug 20]. Available from: https://www.ippc.int/en/core-activities/standards-setting/ispms/

[14] Mahajan P V., Caleb OJ, Gil MI, Izumi H, Colelli G, Watkins CB, et al. Quality and safety of fresh horticultural commodities: Recent advances and future perspectives. Food Package Shelf Life [Internet]. 2017;14(August):2-11. Available from: http://dx.doi.org/10.1016/j.fpsl.2017.08.001

[15] Elhadi M Yahia. Modified and controlled atmospheres for tropical fruits. Stewart Postharvest Rev. 2006;2(5):1-10.

[16] Arvanitoyannis IS, Stratakos AC, Tsarouhas P. Irradiation applications in vegetables and fruits: A review. Crit Rev Food Sci Nutr. 2009;49(5):427-462.

[17] Singh SP, Saini MK. Postharvest vapour heat treatment as a phytosanitary measure influences the aroma volatiles profile of mango fruit. Food Chem [Internet]. 2014;164:387-395. Available from: http://dx.doi.org/10.1016/j.foodchem.2014.05.009

[18] Ware AR, Du Toit CLN, Mohamed SA, Nderitu PW, Ekasi S. Cold tolerance and disinfestation of *Bactrocera invadens* (Diptera: Tephritidae) in "Hass" avocado. J Econ Entomol. 2012;105(6):1963-1970.

[19] Ndlela S, Ekesi S, Ndegwa PN, Ong'amo GO, Mohamed SA. Post-harvest disinfestation of *Bactrocera dorsalis* (Hendel) (Diptera: Tephritidae) in mango using hot-water treatments. J Appl Entomol. 2017;141(10):848-859.

[20] Ocitti P, Ndlela S, Akol A, Muyinza M, Mohamed S. Tephritidae) in Tommy Atkins mango using hot-water immersion treatment. African Entomol. 2021;29(1):238-247.

[21] Jacobi KK, MacRae EA, Hetherington SE. Postharvest heat disinfestation treatments of mango fruit. Sci Hortic (Amsterdam). 2001;89(3):171-193.

[22] Lux SA, Copeland RS, White IM, Manrakhan A, Billah MK. A New Invasive Fruit Fly Species from the *Bactrocera dorsalis* (Hendel) Group Detected in East Africa. Int J Trop Insect Sci. 2003;23(4):355-361.

[23] USDA-APHIS [United States Department of Agriculture- Animal and Plant Health Inspection Service]. Federal import quarantine order for host materials of *Bactrocera invadens* [Diptera: Tephritidae], invasive fruit fly species. 2008 [cited 2021 Aug 20]; Available from: https://www.aphis.usda.gov/aphis/banner/help

[24] EU. Commission Implementing Directive [EU] 2019/523 of 21 March 2019 amending Annexes I to V to Council Directive 2000/29/EC on protective measures against the introduction into the Community of organisms harmful to plants or plant products and against their spread within the Community. Off J Eur Union L [Internet]. 2019 [cited 2021 Aug 20];41-68. Available from: https://eur-lex.europa.eu/legal-content/EN/TXT/?uri=celex:32019L0523

[25] José L, Cugala D, Santos L. Assessment of invasive fruit fly fruit infestation and damage in Cabo Delgado province, northern Mozambique. African Crop Sci J. 2013;21(1):21-28-28.

[26] Cugala D, José L, Mahumane C, Mangana S. Fruit flies pest status, with emphasis on the occurrence of the invasive fruit fly, *Bactrocera invadens*

(Diptera: Tephritidae) in Mozambique. In: African Crop Science Society Conference [Internet]. 2009 [cited 2021 Aug 20]. p. Vol 28. Available from: https://scholar.google.com/scholar?hl=en&as_sdt=0%2C5&q=Fruit+flies+pest+status%2C+with+emphasis+on+the+occurrence+of+the+invasive+fruit+fly%2C+Bactrocera+invadens+%5BDiptera%3A+Tephritidae%5D+in+Mozambique.+In+African+Crop+Science+Society+Conference+200&btnG=#d=gs_cit&u=%2Fscholar%3Fq%3Dinfo%3AOEgkr81HlWkJ%3Ascholar.google.com%2F%26output%3Dcite%26scirp%3D0%26hl%3Den

[27] Otieno W. EPHIS experience with market access and compliance with official standards. In: All Africa Horticultural Congress 911 [Internet]. 2009 [cited 2021 Aug 20]. p. 73-6. Available from: https://www.actahort.org/books/911/911_8.htm

[28] Escribano S, Mitcham EJ. Progress in heat treatments. Stewart Postharvest Rev [Internet]. 2014;10:1-6. Available from: http://www.ingentaconnect.com/content/sphs/sphr/2014/00000010/00000003/art00003

[29] Mutyambai DM, Mbeche NI, Onamu E, Kasina MJ, Nderitu JH, Mweke AN. False codling moth, *Thaumatotibia leucotreta* (Meyrick) a new threat to horticulture industry: Stakeholders' perspectives on the status, impact and management in Kenya. J Plant Dis Prot [Internet]. 2020;127(6):799-804. Available from: https://doi.org/10.1007/s41348-020-00363-5

[30] Daiber CC. study of the biology of the false codling moth (*Cryptophlebia leucortreta* (Meyr.)): the larva. Phytophylactica [Internet]. 1979;157(March):151-157. Available from: http://agris.fao.org/agris-search/search.do?f=2013/US/US2013028198410020843.xml;US201302819908

[31] CABI. *Thaumatotibia leucotreta* (false codling moth (FCM)) In: Invasive Species Compendium [Internet]. Wallingford, UK: CAB International. 2020 [cited 2021 Aug 20]. Available from: https://www.cabi.org/isc/datasheet/6904

[32] Gilligan TM, Epstein ME, Hoffman KM. Discovery of false codling moth, *Thaumatotibia leucotreta* (Meyrick), in California (Lepidoptera: Tortricidae). Proc Entomol Soc Washingt. 2011;113(4):426-435.

[33] Moore SD. Biological control of a phytosanitary pest (*Thaumatotibia leucotreta*): A case study. Int J Environ Res Public Health. 2021;18(3):1-19.

[34] Jacobi KK, Wong LS. Quality of "Kensington" mango (Mangifera indica Linn.) following hot water and vapour-heat treatments. Postharvest Biol Technol. 1992;1(4):349-359.

[35] Lurie S. Postharvest heat treatments. Postharvest Biol Technol [Internet]. 1998 [cited 2021 Aug 20];14(3):257-69. Available from: https://www.sciencedirect.com/science/article/pii/S0925521498000453

[36] Indiarto R, Izzati AN, Djali M. Post-harvest handling technologies of tropical fruits: A review. Int J Emerg Trends Eng Res. 2020;8(7):3951-3957.

[37] Sharp JL. Hot-water Treatment for Control of *Anastrepha suspensa* (Diptera: Tephritidae) in Mangos. J Econ Entomol [Internet]. 1986 Jun 1 [cited 2021 Aug 20];79(3):706-8. Available from: https://academic.oup.com/jee/article/79/3/706/2214662

[38] Hallman GJ, Sharp JL. Mortality of Caribbean Fruit Fly (Diptera: Tephritidae) Larvae Infesting Mangoes Subjected to Hot-Water Treatment, then Immersion Cooling. J Econ Entomol [Internet]. 1990 Dec 1 [cited 2021 Aug 20];83(6):2320-3. Available from:

https://academic.oup.com/jee/article/83/6/2320/871072

[39] Sharp JL, Ouye MT, Ingle SJ, Hart WG. Hot-Water Quarantine Treatment for Mangoes from Mexico Infested with Mexican Fruit Fly and West Indian Fruit Fly (Diptera: Tephritidae). J Econ Entomol. 1989;82(6):1657-1662.

[40] Sharp JL, Ouye MT, Thalman R, Hart W, Ingle S, Chew V. Submersion of 'Francis' Mango in Hot Water as a Quarantine Treatment for the West Indian Fruit Fly and the Caribbean Fruit Fly (Diptera: Tephritidae). J Econ Entomol [Internet]. 1988 Oct 1 [cited 2021 Aug 20];81(5):1431-6. Available from: https://academic.oup.com/jee/article/81/5/1431/2214949

[41] Spalding DH, King JR, Sharp JL. Quality and decay of mangos treated with hot water for quarantine control of fruit fly. Trop Sci [Internet]. 1988 [cited 2021 Aug 14];28(2):99-101. Available from: http://www.sidalc.net/cgi-bin/wxis.exe/?IsisScript=orton.xis&method=post&formato=2&cantidad=1&expresion=mfn=003029

[42] Smith E, Chin D. Hot water dipping as a disinfestation treatment against the fruit fly *Dacus aquilonis* (May)(Diptera: Tephritidae) in mangoes. In: III International Mango Symposium 291 [Internet]. 1989 [cited 2021 Aug 20]. p. 389-403. Available from: https://www.actahort.org/books/291/291_44.htm

[43] Joyce D, Hockings P, Mazucco R, Shorter A, Brereton I. Heat treatment injury of mango fruit revealed by nondestructive magnetic resonance imaging. Postharvest Biol Technol [Internet]. 1993 [cited 2021 Aug 20];3(4):305-11. Available from: https://www.sciencedirect.com/science/article/pii/092552149390011Q

[44] Segarra-Carmona AE, Franqui RA, Ramírez-Ramos L V., Santiago LR, Torres-Rivera CN. Hot water dip treatments to destroy Anastrepha obliqua larvae (Diptera: Tephritidae) in mangoes from Puerto Rico. J Agric Univ Puerto Rico. 1990;74(4):441-447.

[45] Djioua T, Charles F, Lopez-Lauri F, Filgueiras H, Coudret A, Jr MF, et al. Improving the storage of minimally processed mangoes (*Mangifera indica* L.) by hot water treatments. Postharvest Biol Technol. 2009 May 1;52(2):221-226.

[46] Anwar R, Malik AU. Effect of hot water treatment on storage life and quality of mango (*Mangifera indica* L.). Acta Hortic. 2007;768(2):201-207.

[47] McGuire R. Concomitant decay reductions when mangoes are treated with heat to control infestations of Caribbean fruit flies. Plant Dis [Internet]. 1991 [cited 2021 Aug 20];75(9):946-9. Available from: https://pascal-francis.inist.fr/vibad/index.php?action=getRecordDetail&idt=5567337

[48] Verghese A, Sreedevi K, Nagaraju D. Pre and post harvest IPM for the mango fruit fly, *Bactrocera dorsalis* (Hendel). In: 7 International symposium on fruit flies of economic importance: from basic to applied knowledge. Salvador, BA (Brazil): ETDWEB; 2006. p. 10-15.

[49] Nyanjage MO, Wainwright H, Bishop CFH. Effects of hot-water treatment and storage temperature on electrolyte leakage of mangoes (*Mangifera indica* Linn.). J Hortic Sci Biotechnol [Internet]. 1999 [cited 2021 Aug 20];74(5):566-72. Available from: https://www.tandfonline.com/doi/abs/10.1080/14620316.1999.11511154

[50] Hofman PJ, Stubbings BA, Adkins MF, Meiburg GF, Woolf AB. Hot water treatments improve 'Hass' avocado fruit quality after cold disinfestation. Postharvest Biol Technol. 2002 Mar 1;24(2):183-192.

[51] Woolf AB, Laing WA. Avocado Fruit Skin Fluorescence following Hot Water

Treatments and Pretreatments. J Am Soc Hortic Sci [Internet]. 1996 Jan 1 [cited 2021 Aug 20];121(1):147-51. Available from: https://journals.ashs.org/jashs/view/journals/jashs/121/1/article-p147.xml

[52] Woolf AB, Lay-Yee M. Pretreatments at 38 °C of `Hass' Avocado Confer Thermotolerance to 50 °C Hot Water Treatments. HortScience [Internet]. 1997 Jul 1 [cited 2021 Aug 20];32(4):705-8. Available from: https://journals.ashs.org/hortsci/view/journals/hortsci/32/4/article-p705.xml

[53] González-Aguilar GA, Cruz R, Baez R, Wang CY. Storage quality of bell peppers pretreated with hot water and polyethylene packaging. J Food Qual. 1999;22(3):287-299.

[54] González-Aguilar GA, Gayosso L, Cruz R, Fortiz J, Báez R, Wang CY. Polyamines induced by hot water treatments reduce chilling injury and decay in pepper fruit. Postharvest Biol Technol. 2000;18(1):19-26.

[55] Fallik E, Grinberg S, Alkalai S, Yekutieli O, Wiseblum A, Regev R, et al. A unique rapid hot water treatment to improve storage quality of sweet pepper. Postharvest Biol Technol. 1999 Jan 1;15(1):25-32.

[56] Sakaldas M, Kaynas K. Biochemical and quality parameters changes of green sweet bell peppers as affected by different postharvest treatments. African J Biotechnol [Internet]. 2010 [cited 2021 Aug 20];9(48):8174-81. Available from: https://www.ajol.info/index.php/ajb/article/view/127356

[57] López-Velázquez JG, Delgado-Vargas F, López-Ángulo G, García-Armenta E, López-López ME, Ayón-Reyna LE, et al. Phenolic profile associated with chilling tolerance induced by the application of a hot water treatment in bell pepper fruit. J Food Sci [Internet]. 2020 Jul 1 [cited 2021 Aug 20];85(7):2080-9. Available from: https://onlinelibrary.wiley.com/doi/full/10.1111/1750-3841.15310

[58] Ilić ZS, Trajković R, Pavlović R, Alkalai-Tuvia S, Perzelan Y, Fallik E. Effect of heat treatment and individual shrink packaging on quality and nutritional value of bell pepper stored at suboptimal temperature. Int J Food Sci Technol. 2012 Jan;47(1):83-90.

[59] Erkan M, Pekmezci M, Wang CY. Hot water and curing treatments reduce chilling injury and maintain post-harvest quality of "Valencia" oranges. Int J Food Sci Technol. 2005 Jan;40(1):91-96.

[60] Connon RE, Geist J, Werner I. Effect-based tools for monitoring and predicting the ecotoxicological effects of chemicals in the aquatic environment. Sensors. 2012;12(9): 12741-12771.

[61] Olea N, Fernandez M. Chemicals in the environment and human male fertility. Occup Environ Med [Internet]. 2007 [cited 2021 Aug 20];64(7):430-1. Available from: https://oem.bmj.com/content/64/7/430.short

[62] Li X, Zhu X, Zhao N, Fu D, Li J, Chen W, et al. Effects of hot water treatment on anthracnose disease in papaya fruit and its possible mechanism. Postharvest Biol Technol [Internet]. 2013;86:437-446. Available from: http://dx.doi.org/10.1016/j.postharvbio.2013.07.037

[63] Vilaplana R, Chicaiza G, Vaca C, Valencia-Chamorro S. Combination of hot water treatment and chitosan coating to control anthracnose in papaya (*Carica papaya* L.) during the postharvest period. Crop Prot. 2020;128(May 2019).

[64] Couey H. Heat treatment for control of postharvest diseases and insect pests of fruits. HortScience [Internet]. 1989

[cited 2021 Aug 20];24(2):198-202. Available from: https://pascal-francis.inist.fr/vibad/index.php?action=getRecordDetail&idt=6718223

[65] USDA-APHIS. Treatment manual. United States Dep Agric [Internet]. 2021;920. Available from: http://www.aphis.usda.gov

[66] Fallik E, Ilić Z. Hot Water Treatments. In: Novel Postharvest Treatments of Fresh Produce [Internet]. 1st Edition. CRC Press; 2017 [cited 2021 Aug 20]. p. 241-58. Available from: https://www.taylorfrancis.com/chapters/edit/10.1201/9781315370149-9/hot-water-treatments-elazar-fallik-zoran-ilić

[67] Kumah P, Appiah F. Effect of hot water treatment on quality and shelf-life of Keitt mango. Agric Biol J North Am. 2011;2(5):806-817.

[68] Self G, Ducamp MN, Vayssières JF. The effects of phytosanitary hot water treatments on west African mangoes infested with *Bactrocera invadens* (Diptera: Tephritidae). Fruits. 2012;67(6):439-449.

[69] Blakey RJ, Bower JP. The feasibility of a hot water treatment for South African avocados (*Persea americana* [Mill.] cv Hass). South African Avocado Grow Assoc Yearb 30. 2007;30:66-8.

[70] Kremer-Köhne S. Hot water treatment of avocado fruit to induce cold tolerance. South African Avocado Grow Assoc Yearb 1999. 1999;22:48-50.

[71] Kritzinger M, Kruger FJ, Bezuidenhout M. Further Evaluation of Hot Water / Air Heatshock Treatment of South African Avocados. South African Avocado Grow Assoc Yearb 1998. 1998;21:93-6.

[72] Setagane L, Mafeo T, Mathaba N, Shikwambana K. Mitigation of chilling injury with hot water treatment to improve early-season 'HASS'avocado (*Persea americana*) fruit peel colour. Res Crop [Internet]. 2021 [cited 2021 Aug 20];22(1). Available from: http://search.ebscohost.com/login.aspx?direct=true&profile=ehost&scope=site&authtype=crawler&jrnl=09723226&AN=149856641&h=G1HUqooL5Ks9IARDdYGuV%2Bw9ktM9S%2B3qp%2BNaYgCLQvuHVsTMC5ehmz8VoeSdv9GqnIgzf0ukNM28jPO3IpuD9g%3D%3D&crl=c

[73] Kassim A, Workneh TS. Influence of postharvest treatments and storage conditions on the quality of Hass avocados. Heliyon. 2020 Jun 1;6(6):e04234.

[74] Shehata SA, Ibrahim ; M I A, El-Mogy MM, Abd El-Gawad KF. Effect of hot water dips and Modified Atmosphere packaging on extend the shelf life of bell pepper fruits. J Klagefurt, Austria. 2013;20(3):315-328.

[75] Speckhahn C, Subramanian S, Meyhofer R. Postharvest warm water treatment to control thrips in export French beans. In: The 1st All Africa Post Harvest Congress & Exhibition, Reducing food losses and waste: sustainable solutions for Africa, 28th-31st March 2017, Nairobi, Kenya Conference Proceedings. Nairobi, Kenya: University of Nairobi; 2017. p. 7-9.

[76] FAOSTAT. Crops. Food and Agriculture Organization of the United Nations [Internet]. FAO. 2020 [cited 2021 Aug 20]. Available from: http://www.fao.org/faostat/en/#data/QC

[77] Dohino T, Hallman GJ, Grout TG, Clarke AR, Follett PA, Cugala DR, et al. Phytosanitary Treatments against *Bactrocera dorsalis* (Diptera: Tephritidae): Current Situation and Future Prospects. J Econ Entomol. 2016;110(1):67-79.

[78] Mohamed A, Kamal I, Rauof F, Hussein D, Babiker S, Elsheikh B, et al.

Effectiveness and suitability of vapor heat treatment in disinfestation of export mango fruit, cultivar Abu Samaka, from fruit flies. Gezira J Agric Sci [Internet]. 2017 Jun 1 [cited 2021 Aug 20];15(1). Available from: http://journals.uofg.edu.sd/index.php/gjas/article/view/93

[79] Sikuka W. Strong Domestic and Export Demand Drives Growth in South African Avocado Plantings | USDA Foreign Agricultural Service [Internet]. Pretoria, South Africa; 2021 Feb [cited 2021 Aug 20]. Available from: https://www.fas.usda.gov/data/south-africa-strong-domestic-and-export-demand-drives-growth-south-african-avocado-plantings

[80] Kok R, Bower JP, Bertling I. Low temperature shipping and cold chain management of 'Hass' avocados: An opportunity to reduce shipping costs RD. South African Avocado Grow Assoc Yearb 33. 2010;33:33-7.

[81] Kassim A, Workneh T, Bezuidenhout C. A review on postharvest handling of avocado fruit. African J Agric Res [Internet]. 2013 [cited 2021 Aug 20];8(21):2385-402. Available from: https://academicjournals.org/journal/AJAR/article-abstract/E6F076B34025

[82] Jessup J. Curing "Hass" avocados for cold storage dismfestation against Queensland Fruit Fly AV010. 1993.

[83] Jensen SE. Insecticide Resistance in the Western Flower Thrips, *Frankliniella occidentalis.* Integr Pest Manag Rev 2000 52 [Internet]. 2000 [cited 2021 Aug 20];5(2):131-46. Available from: https://link.springer.com/article/10.1023/A:1009600426262

[84] Przybylska A, Fiedler Z, Obrępalska-Stęplowska A. PCR-RFLP method to distinguish *Frankliniella occidentalis*, *Frankliniella intonsa*, *Frankliniella pallida* and Frankliniella tenuicornis. J Plant Prot Res. 2016;56(1):60-66.

[85] Armstrong J, Mangan RL. Commercial quarantine heat treatments. In: Tang J, Mitcham E, Wang S, Lurie S, editors. Heat Treatments for Postharvest Pest Control: Theory and Practice [Internet]. Wallingford, UK: CAB International; 2007 [cited 2021 Aug 20]. p. 311-40. Available from: https://books.google.com/books?hl=en&lr=&id=YnyvsYhr5-gC&oi=fnd&pg=PA311&dq=Commercial+quarantine+heat+treatments.+Heat+Treatments+for+Postharvest+Pest+Control:+In+Tang+J,+Mitcham+E,+Wang+S,+and+Lurie+&ots=Zh6WRvulE-&sig=MyRSOYAWLk53HtrDQiIqT8sx7pM

[86] Jing W, Tu K, Shao XF, Su ZP, Zhao Y, Wang S, et al. Effect of postharvest short hot-water rinsing and brushing treatment on decay and quality of strawberry fruit. J Food Qual. 2010;33(SUPPL. 1):262-272.

[87] Porat R, Daus A, Weiss B, Cohen L, Fallik E, Droby S. Reduction of postharvest decay in organic citrus fruit by a short hot water brushing treatment. Postharvest Biol Technol. 2000;18(2):151-157.

[88] Irtwange S. Hot Water treatment: A Non-Chemical Alternative in Keeping Quality During Postharvest Handling of Citrus fruits. Agric Eng Int CIGR Ejournal. 2006;8(5):1-10.

[89] Jordan RA. The disinfestation heat treatment process. Plant quarantine in Asia and the Pacific. In: A Report of an Asian Productivity Organization Study Meeting, Taipei, Taiwan, 17-26 March 1992 [Internet]. Tokyo: Asian Productivity Organization; 1993 [cited 2021 Sep 7]. p. 53-68. Available from: https://scholar.google.com/scholar?hl=en&as_sdt=0%2C5&q=The+disinfestation+heat+treatment+process.+Plant+quarantine+in+Asia+and+the+Pacific&btn

G=#d=gs_cit&u=%2Fscholar%3Fq%3Dinfo%3ATfU4F9EBd1cJ%3Ascholar.google.com%2F%26output%3Dcite%26scirp%3D0%26hl%3Den

[90] Amin MN, Hossain MM. Reduction of Postharvest Loss and Prolong the Shelf-Life of Banana through Hot Water Treatment. J Chem Eng. 2012;27(1):42-47.

[91] Nguyen HT, Boonyaritthongchai P, Buanong M, Supapvanich S, Wongs-Aree C. Postharvest Hot Water Treatment Followed by Chitosan- and κ-Carrageenan-Based Composite Coating Induces the Disease Resistance and Preserves the Quality in Dragon Fruit (*Hylocereus undatus*). Int J Fruit Sci [Internet]. 2020;20(S3):S2030-44. Available from: https://doi.org/10.1080/15538362.2020.1851342

Section 4

Postharvest Processing and Packaging

Chapter 7

Advances in Postharvest Packaging Systems of Fruits and Vegetable

Trina Adhikary and Durga Hemanth Kumar

Abstract

The production of vegetables and fruits is at a high rate but the major challenging task is the postharvest handling and processing of the products. Approximately 20–30% of the production is being wasted due to a lack of proper postharvest management. Many developments were made to reduce this wastage such as cold chain development, different storage structures, some drying methodologies to promote the shelf life of produce. But all these systems need to be improved and utilized commercially. The losses still occur due to a lack of sound knowledge on the chemical nature of products and different management techniques (e.g., drying, cooling, blanching). Therefore, the successful design of the cooling, packing, storage transport, and drying processes of fresh food requires linking materials sciences, fluid dynamics, mechanical deformation, food chemistry, and process control.

Keywords: packing, advanced packing systems, bio-degradable packing, shelf life

1. Introduction

Fruits and vegetables are highly perishable and have a very short shelf-life. During different handling and marketing operations, there is a huge postharvest loss of agricultural produce. Both qualitative and quantitative losses occur in horticultural commodities between harvest and consumption. Qualitative losses like loss inedibility, nutritional quality, calorific value, and consumer acceptability of fresh produce are much more difficult to assess than are quantitative losses [1]. Quantitative post-harvest losses in India estimated by different committees ranged between 25 and 33% depending upon the crop. The major cause of postharvest loss is the lack of proper infrastructure for processing and packing. These losses can only be minimized to some extent by proper marketing, handling, and processing of agricultural commodities. According to a national level study conducted under the All India Coordinated Research Project (AICRP) on Postharvest technology of the Indian Council of Agricultural Research (ICAR) the post-harvest losses during different farm handling operations like harvesting, sorting, grading, and packing accounts for about 13%, during farm storage about 6% and during storage at going down, wholesale and retail level about 12% of the produce goes waste. Thus, on average, about one-third of horticulture produce never reaches the ultimate consumer. This results in a considerable gap between gross food production and net availability [2]. Insufficient knowledge of pre and post-harvest operations and lack of proper facilities for handling like pre-cooling, grading, packaging, transport,

IntechOpen

storage, processing, and marketing all together compound the post-harvest losses and wastage which in value terms accounts for more than 6,720,000.00 US dollars.

Keeping the huge postharvest losses in mind, there is an urgent need to reduce the postharvest losses of fresh commodities and increase the level of processing as a reduction in post-harvest losses is a complementary means of production [3]. The important strategies for loss prevention include the development of varieties (genotypes) that have longer postharvest life, use of integrated crop management system, and development of cost-effective adaptable technologies for post-harvest handling, value addition, and by-product waste utilization [4]. The value chain in post-harvest management of horticultural crops mainly comprises pre-harvest factors, harvesting, market preparation (pre-cooling, sorting, grading, packaging, and on-farm storage), transportation, storage, value addition, and by-product waste management. The status of R&D carried out pertaining to postharvest management (PHM) and processing in the country by different ICAR institutes like Central Institute of Post Harvest Technology (CIPHET) (Ludhiana) and State Agricultural University (SAUs) on different aspects of post-harvest management and processing of horticultural crops is given ahead. Depending upon the status report, research scientists can find out the gap/missing links in the available technology to suggest future priorities in the area of R&D.

Maturity is the state where the product is ready for picking. Proper identification of maturity of produce is essential so that the product is less prone to various physiological disorders and diseases [5]. Maturity indices have been developed for various fruits such as mango, pomegranate, apple, grapes, ber, aonla, Nagpur mandarin, etc. Technique to determine the maturity of mango on the tree (CIPHET) and non-destructive method for the maturity of Grand Naine banana (NRCB, Trichur) need to be popularized.

In recent years, rapid industrialization, population growth, and changed lifestyle led to increased demand for processed and packed foods. Currently, ready to eat packed food industry is growing very fast. Packaging is considered as the science, art, and technology of protecting the products during transportation, distribution, storage, sale, and use. Further, the packaging ensures safe and efficient delivery of the commodity to the consumer in good condition. Good packaging attracts the customer to buy the product. It also plays a vital role in reducing the security risks during shipment. Packaged products are easy in displaying, handling, storing, distributing, opening, reclosing, and reusing. Packaging performs four important functions, such as containment, protection, convenience, and communication. A wide variety of materials, such as cane baskets, wooden boxes, clay vessels, metal cans, China pots, paper bags, and plastics containers are still used for packaging the products in many areas of the world. The packaging material should not cause any environmental pollution. Hence, there is a need to undertake detailed studies to assess the impact of food packaging on the environment.

In this context, Paine and Paine [6] concluded that packaging contains, protects, and preserves as well as informs to create convenience to consumers. It is stated that many companies apply packaging to create values beyond the basic components of containing, protecting, preserving, and informing [7]. Recent progress in food packaging is resulting from the rising need for mild processed but with better shelf-life food products by the consumers. An important reason for innovative packaging is the emergence of food-borne microbial outbreaks that demand packaging with anti-microbial products to ascertain quality and safety. No hazardous components must touch the food within the packaging, and the flavor of the food should not get affected. The food must not change its original appearance and taste. In addition, the food should not cause any discoloring in the packaging. It is pertinent to mention that high-quality films serve to protect a product during transportation, distribution, and use. It seems that the public health impact of unhygienic packaging of food is not well studied. The new food packaging techniques, such as intelligent packaging, bio-active packaging, and active packaging, which engage deliberate contact with the food or its surroundings and influence on consumer's health have been the most important innovations in the field of packaging technology [8]. Therefore, the main objective of this article is to present an overview of the innovations in food packaging technology.

2. Functions of packaging

It is essential to minimize physical damage to fresh produce to obtain optimal shelf-life. The use of suitable packaging is vital in this respect [9]. The most frequently used one is the fibreboard carton, however, they may vary depending on the product and its physical nature, for example, tissue paper wraps, trays, cups or pads, are required to reduce damage from abrasion. Individual packing of the product is most suitable as it ensures its microenvironment and also reduces physical contact with others which improves its texture and nature and prevents the spread of disease-causing pathogens. Molded trays may be used which physically separate the individual piece of produce. Packing plays a crucial role in enhancing the post-harvest life of produce and ideal packing material should possess some characters:

- Readily available
- Easy to handle i.e., less weight
- Cost-effective
- Provide adequate ventilation for produce
- Eco friendly

When packaging is required at the source or when an extended storage life is desired, the packaging film should have high gas permeability and anti-fog properties.

The most commonly used packing material at local markets or for retail purposes is polyethylene (PE) bags. The packaging of fresh vegetables and fruit provides the largest single use of printed PE bags. But they do not have their presence in long-distance transport as they are not firm enough and may cause destruction to the product that results in decay and economic loss to the marketer. During packing the principal factor to be taken into consideration is free movement of air so that the temperature within the enclosure does not increase and shelf life is not affected. Light does not seem to be an essential factor for packing, however, some green leafy vegetables perform photosynthesis by absorbing carbon dioxide and release oxygen upon exposure to light. Vibrations and shock may cause damage to cells that leads to increased respiration rate and enzymes to be released that cause browning reaction to getting started.

3. Requirements of efficient food packaging process

The important requirements of food packages are given as follows (ICAR online e-courses).

- It should protect from physical damage.

- It should safeguard from contamination.
- It should protect from bad smells and external toxicants.
- It should be nontoxic.
- It should not affect the food packaging.
- It should be easy to open.
- It should act as a barrier for moisture and oxygen ingress.
- It should filter harmful ultraviolet light.
- It should meet the required physical requirements.
- It should be transparent and resistant or tamper.
- It should have appearance and printability features.
- It should be of low cost.
- I should have handling features.
- It should be disposed of easily.

4. Different types of packing systems

4.1 Modified atmosphere packaging (MAP)

Polymeric films are regularly used because of their advantages and their availability, the chief factor in their control of movement and concentration of gasses by lowering the oxygen concentration and raising carbon dioxide concentration that abridges the respiration rate and promotes produce shelf-life (controlled atmospheric (CA) packing). Temperature control plays a crucial role in modified atmosphere packaging (MAP) packing as it directly influences respiration rate that shows an effect on the shelf life of produce. The major drawback of MAP packing is that the concentration of O_2 is reduced to a greater extent that may result in the fermentation of tissues producing undesirable off-flavors.

MAP can be done in 2 ways:

1. **Active:** it involves creating a vacuum within the product and replacing it with desired gaseous concentration. Some absorbers may also be used to control gas concentration (**Tables 1** and **2**).

2. **Passive:** the atmosphere within the product is attained because its respiration, final equilibrium depends on the characters of the commodity.

However, the packing material used may not satisfy all the properties required, so they are combined to provide a wide range of characters by lamination and co-extrusion. The concentration of gasses accumulated depends on many variables

DOI: http://dx.doi.org/10.5772/intechopen.101124

Film	Permeability (cm^3/m^2 day atm) for 25 μm film at 25°C			WTR (g/m^2/day/atm) at 38°C, 90% RH
	O_2	N_2	CO_2	
Ethylene-vinyl alcohol (EVAL)	3–5	—	—	16–18
Polyvinylidenechloride (PVdC)-PVC copolymer (Saran)	9–15	—	20–30	—
Low-density polythene (PE-LD)	7800	2800	42,000	18
High-density polyethylene (PE-HD)	2600	650	7600	7–10
Polypropylene cast (PPcast)	3700	680	10,000	10–12
Polypropylene, oriented (OPP)	250,000	400	8000	6–7
Polypropylene, oriented, PVdC coated (OPP/PVdC)	10–20	8–13	35–50	4–5
Rigid poly (vinyl chloride) PVC	150–350	60–150	450–1000	30–40
Plasticized poly(vinyl chloride) (PVC-P)	500–30,000	300–10,000	1500–46,000	15–40
Ethylene-vinyl acetate (EVAC)	12,500	4900	50,000	40–60
Polystyrene, oriented (OPS)	5000	800	18,000	100–125
Polyurethane (PUR)	800–1500	600–1200	7000–25,000	400–600
PVdC-PVC copolymer (Saran)	8–25	2–2.6	50–150	1.5–5.0
Polyamide (Nylon-6), (PA)	40	14	150–190	84–3100

[10–15].

Table 1.
Gas permeability and water transmission rate (WTR) of polymeric film available for packaging of MAP produce.

Fruits	O_2 (%)	CO_2 (%)	N_2 (%)	Vegetables	O_2 (%)	CO_2 (%)	N_2 (%)
Apple	1–2	1–3	95–98	Artichoke	2–3	2–3	94–96
Apricot	2–3	2–3	94–96	Beans, snap	2–3	5–10	87–93
Avocado	2–5	3–10	85–95	Broccoli	1–2	5–10	88–94
Banana	2–5	2–5	90–96	Brussels sprouts	1–2	5–7	91–94
Grape	2–5	1–3	92–97	Cabbage	2–3	3–6	81–95
Grapefruit	3–10	5–10	80–92	Carrot	5	3–4	91–95
Kiwifruit	1–2	3–5	93–96	Cauliflower	2–5	2–5	90–96
Lemon	5–10	0–10	80–95	Chili peppers	3	5	92
Mango	3–7	5–8	85–92	Corn, sweet	2–4	10–20	76–88
Orange	5–10	0–5	85–95	Cucumber	3–5	0	95–97
Papaya	2–5	5–8	87–93	Lettuce (leaf)	1–3	0	97–99
Peach	1–2	3–5	93–96	Mushrooms	3–21	5–15	65–92
Pear	2–3	0–1	96–98	Spinach	Air	10–20	—
Pineapple	2–5	5–10	85–93	Tomatoes	3–5	0	95–97
Strawberry	5–10	15–20	70–80	Onion	1–2	0	98–99

[2, 11, 16].

Table 2.
Recommended gas mixtures for MAP.

such as the chemical composition of products, packing material permeability, product respiration, and the influence of temperature on them. A lot of commercial interest has been focused on developing packing materials with high gas transmission rates. For major polythene films have more permeability to CO_2 than O_2, thus aid in maintaining a proper gaseous ratio. Thus, packaging film of the correct permeability must be chosen to realize the full benefits of MAP of fresh produce [17].

Typical packing material should have a 2–10% O_2/CO_2 ratio to maintain the freshness of produce and enhance its shelf life. Highly respiring produce must not be loaded in traditional packing material such as poly(vinyl chloride) (PVC), low-density polythene (PE-LD), polypropylene, oriented (OPP), instead kept in the highly permeable micro-perforated film so that the gaseous concentration is maintained. Ceramic films have high oxygen, carbon dioxide, ethylene permeability [18]. Films that have high gas permeability are usually a mixture of two or more non-numeric units each contributing a specific character such as strength, transmission, durability, permeability, etc. Furthermore, films can be laminated to achieve desired traits Films using micro-perforations can attain very high rates of gas transmission [19]. Films with micro-perforations are preferred, generally, the size ranges from 40 to 200 μm, and by making modifications to them we can regulate the gaseous concentration to meet product requirements. Based on the release of gasses from perforations of film, suitable packing materials have been identified for mushrooms. Perforated packing materials also proved good to store nectarines, apples, asparagus, etc. Macro perforated material can also be used to pack some strawberries and raspberries. Micro-perforated material is expensive and may also allow entry of some pathogens during wet handling conditions [17].

The most effective and efficient way for packing high respiring produce is by combining high O_2 MAP and low O_2 MAP, because of high oxygen concentration there is the prevention of off-flavors and odd odors that result due to fermentation [11, 17]. Macro perforated material can also be used to pack some strawberries and raspberries. Micro-perforated material is expensive and may also allow entry of some pathogens during wet handling conditions [17].

The most effective and efficient way for packing high respiring produce is by combining high O_2 MAP and low O_2 MAP, because of high oxygen concentration there is the prevention of off-flavors and odd odors that result due to fermentation [11, 17].

- Proper movement of air must be ensured for enhancing the shelf life of produce and also increase resistance to gas diffusion. Ethylene is known as a natural ripening hormone and is active at trace concentrations, it is observed that its activity is reduced at oxygen levels of 2–10%, thus low oxygen enhances shelf life.

- Biological reactions increase by 2–3 times for every 10°C rises in temperature, film permeability also increases with fluctuations in temperature hence temperature control is crucial for successful MAP, temperature fluctuations may result in browning of tissues, loss of firmness, increased ethanol content, all in combination deteriorate the quality of produce packed.

- Relative humidity (RH) also has to affect produce packed, more RH invites disease-causing pathogens thus reduces the quality of produce, whereas low RH increases transpiration damage and leads to desiccation. A mathematical model was developed for estimating the changes in the atmosphere and humidity within perforated packages of fresh produce [18, 20, 21]. This model depends on the concentrations of O_2, CO_2, N_2, and H_2O vapors in the package. A different procedure was developed to maintain the concentrations of O_2 and CO_2 inside packages that are exposed to different environmental conditions [22].

- Cucumbers that are not packaged experienced severe chilling injury compared to those packed in 31.75 μm PE-LD when they are stored at 5°C and 90–95% RH [3]. The influence of MAP on the sensory characteristics and shelf life of shiitake mushrooms (*Lentinula edodes*) was also studied using PE-LD, polypropylene (PP), and macro perforated film.

- Some fresh vegetable shelf life has been enhanced by packing them with nitrogen gas.

4.2 Edible coatings and films

Increased use of synthetic packing material poses an environmental threat during its disposal, hence some coating techniques evolved that satisfy both the product shelf life and less threat to nature (**Table 3**). The materials used or coating must full fill some features such as acceptable sensorial characteristics, appropriate barrier properties, good mechanical strength, reasonable microbial, biochemical, and physicochemical stability, safety, low cost, and simple technology for their production [23].

Mostly used coating materials are polysaccharides of starch, proteins, the cellulose that does not pose any harm to human health. Carboxymethylcellulose is one of the materials that gained attention because of its wide applications. The materials used may be extracted from plants such as (corn zein, wheat protein, soy protein) or from animals (casein, whey protein). Pullulan, produced by *Aureobasidium pullulans*, is capable of forming edible films but it is been largely exploited as a coating material, because of its high water solubility. One example of pullulan used as a coating hydrocolloid was for strawberries and kiwifruit [23].

Film	Thickness (mm)	Permeability at 0% RH (10^{-15} l/m^2 s Pa)		Permeability ratio (CO_2/O_2)
		O_2	CO_2	
Corn-zein	0.12–0.31	0.36 30°C	2.67 21°C	7.5
Wheat gluten	0.23–0.42	0.20 30°C	2.13 21°C	9.5
Methyl cellulose low level (MC (L))	0.04–0.07	2.17 30°C	69.00 21°C	31.6
Hydroxypropylcellulose low level (HPC (L))	0.05	3.57 30°C	143.99 21°C	40.6
HPC/lipids	0.15	3.44 30°C	81.75 21°C	23.7
Cozeen	0.09	0.89 37.8°C	5.25 22.8°C	5.9
Wheat gluten	0.14	0.09 37.8°C	0.03 22.8°C	0.3
Corn-zein	0.08	0.16 25°C	—	—
Wheat gluten	0.15	0.08 25°C	—	—

[10].

Table 3.
Oxygen and carbon dioxide permeabilities of edible films.

4.3 Antimicrobial packaging

It's the combination of edible packing material with some antimicrobial agents that aid in inhibiting the growth of microbes. There are several categories of antimicrobials that include, organic acids (acetic, benzoic, lactic, propionic, sorbic), fatty acid esters (glyceryl monolaurate), polypeptides (lysozyme, peroxidase, lactoferrin, nisin), plant essential oils (cinnamon, oregano, lemongrass), nitrites and sulfites, among others [24]. But their use is limited in fresh-cut fruits, only organic acids, and plant essential oils are used. The drawback is that fruits are losing their natural flavor and aroma due to the usage of essential oils. To confer antimicrobial activity, antimicrobial agents may be coated, incorporated, immobilized, or surface modified onto package materials [25].

Antimicrobial films are of 2 types: (a) mobile-which includes an antimicrobial agent that migrates on the surface of produce and prevents pathogenic growth (b) static that does not migrate and inhibits pathogen growth on the surface of produce. Packing materials with grapefruit seed extract in combination with a polyamide binder had an impact on microbial activity compared to grapefruit seed extract (GFSE) alone. When only GFSE is used it should antimicrobial activity against few microbes, but when used in association with a binder it is found effective against several microbes. But these when used alone may not be much effective, hence must be combined with other techniques such as pulsed light, high pressure, and irradiation could reduce the risk of pathogen contamination and extend the shelf-life of perishable food products.

4.4 Active packaging

It is the most efficient technique for packing products that had a dual purpose of maintaining quality and also reduced pathogen damage. It is based on the technique of modifying the internal gas environment by removing or adding gasses to the headspace inside the package. It is done through various ways such as:

- Ethylene scavenging: ethylene is known as a ripening hormone and in very minor concentrations it shows its action, so by eliminating ethylene from packing material we can avoid the further maturation of produce and prevent enzyme action that results in extended shelf life.

- Oxygen scavenging: the presence of oxygen enhances aerobic microbial growth and also enzymatic action. It also results in nutrient loss, off flavor development. Mostly it is used to check mold growth.

- Carbon dioxide release: higher concentrations of carbon dioxide check microbial growth, hence it is essential to maintain it at the needed level, and it is more permeable to plastic films than oxygen, so it must be regulated timely to get quality produce.

- Sulfur dioxide: most commonly used for the packing of grapes, grapes packed in the carton are intermittently fumigated with sulfur dioxide, it must be properly regulated to prevent excess accumulation of sulfur dioxide. Flexible packaging materials such as PE-LD and linear low-density polyethylene (PE-LLD) when impregnated with potassium permanganate and cinnamic acid, respectively become ethylene scavengers.

4.5 Biodegradable packaging

Many biobased polymers are available in the market, like certain kinds of polyester, polyvinyl alcohol, polyesteramides, which are mainly used as films or moldings (**Table 4**). Polyhydroxy acid is very expensive as it is produced in limited quantities at the commercial level. Polylactic acid (PLA) is gaining importance in recent times as it performed better than many synthetic ones. There is always a great demand in searching for biodegradable packing material that serves the dual purpose of being ecofriendly and also less damage to the products stored in it.

The preference of these bio-based packing materials is for those products that need short time storage such as fruits and vegetables. To achieving in this platform the packing material must meet the quality and safety standards of products and also promote its shelf life and fetch good market price to justify the additional costs incurred.

4.6 Application of nanocomposites

They are the nanoscale structures the improve the macro properties of food. Some of the nanocomposites used are silica nano clay and polymer clay nano clay. Silver nano clay have good interactions with other particles and also provides a large surface area to volume ratio, enhanced bacterial activity control, whereas polymer nano clay provide more strength and stiffness, smaller cell size, and is a flame retardant.

Polymer nano clay has recently emerged due to its wide-ranging properties such as providing mechanical strength, less shocking treatments, etc. The properties of biopolymer-based coatings were shown to act as hurdles for gas and solutes thereby increasing the shelf life of produce. But they showed poor performance in mechanical resistance and water vapor exchange. To achieve these characters hybrid materials were developed consisting of bio-based polymer and layered silicates such as montmorillonite (MMT). These exhibited great and good results in the chemical, physical and physiological aspects of the product in comparison to the pure one [27].

Nanocomposite constituents are composed of a nanoscale structure that enhances the macroscopic properties of food products. Polymer clay nano clay and silica nanocomposites of nanosilver are the two common nanocomposites utilized in the food packaging industry. Increased stiffness, strength, nucleating agent in foams, smaller cell size, higher cell density, and flame retardant are the impacts of nano clay in polymers. Nanosilver has great antibacterial characteristics which are made out of de-ionized water suspended in silver. Silver nanoparticles have a large surface area relative to volume, so, they interact well with other particles, increasing their antibacterial efficiency. As a result, they are widely utilized in the food

Material	Film preparation	Moisture barrier	Oxygen barrier	Mechanical properties
Starch/polyvinyl alcohol (PVAL)	Extrusion	–	+	+
Polyhydroxybutyrate/valerate (PHB/V)	Extrusion	+	+	+/–
Polylactic acid (PLA)	Extrusion	+/–	–	+
[26].				

Table 4.
Properties of some biodegradable plastics [26].

packaging business. Although the application of nanotechnology in the food industry was initiated later than other industries, many nanoscientists and technologists have recognized the immense potential of food nanotechnology, particularly in the areas of increasing food quality and ensuring food safety [4].

Polymer/clay nanocomposites are one of the potential applications of nanotechnology in food packaging; they have recently emerged due to their capacity for improving mechanical, barrier, and chemical properties of packaging materials with a small amount of nano clays reinforcement (less than 5% by weight). However major work done on clay polymers concentrated on synthetic polymers majorly. Biopolymers act as a hurdle to solute and gas thereby enhancing the shelf life of produce However, due to their hydrophilic qualities, these films do not retain good mechanical and water vapor barrier capabilities. To overcome these issues, an innovative approach has been developed, by using hybrid materials consisting of polymers and layered silicates such as montmorillonite (MMT) clay mineral, result from the stacked arrangement of negatively charged silicate layers and contain a platelet thickness of about 1 nm with a high aspect ratio (ratio of length to thickness) [28]. The layered silicate filled polymer composites exhibit extraordinary enhancement of mechanical, thermal, and physicochemical properties at a low level of filler concentration when compared to pure polymer and conventional micro composites [27].

In specific, these nanocomposites offer good barrier characteristics, because, the presence of clay layers inhibits the diffusing molecule pathway due to tortuosity [29, 30]. Some of the works done with biopolymer-based nanocomposites were based on starch or polysaccharides, such as chitosan [31, 32], thermoplastic starch and wheat and maize starch. A few studies on protein-based nanocomposites have been available, including whey protein soy protein [31], and wheat gluten. Nanocomposites along with biopolymers exhibited a greater impact when compared nanocomposites alone. The most popular biopolymer is whey protein that has gained popularity due to its transparent coating and effective oxygen barrier. Unlike chitosan film, whey protein films have not shown any antimicrobial activity; therefore, incorporation of antimicrobial agents, such as lysozyme, sorbic acid, and p-aminobenzoic acid and is desirable to induce this feature. Rhim et al., reported that cloister 30R and some chitosan-based nanocomposites showed action against gram-positive bacteria.

4.7 Smart or intelligent packaging

It is of two types: the one which incorporates integrated circuits and the one that does not (chipless smart packing). The type of packing that includes diagnostic indicators also falls under this umbrella. They can be used for some functions such as humidity, light, heat, mechanical shock, biological agents such as bacteria or viruses as they come in contact.

The conventional packing material use Is limited to only some fresh produce and it can not come up with tolerating the high rates of respiration of fresh produce, however, some breathable polymer films were in use for cut vegetables and fruits. Packing films with acrylic side chains is more beneficial as the side chains melt which results in increased gas permeability and also ensures proper carbon dioxide to oxygen ratio that usually varies with the product. In this way, packing becomes smart as the concentrations of gasses are controlled automatically around the product during storage and transportation and provide the products with high quality to the consumers.

Intelligent packaging technique indicates the freshness of produce by changing colors, so the consumer can know its quality and can check it if any deterioration occurred during the transit. Time-temperature integrators (TTI's) are instruments that display irreversible changes in characters such as shape or color. They work based on different principles such as physical, chemical, and biological. The first two types are based on the response towards time, temperature, melting, polymerization, etc... The latter depends on the activity of biological organisms.

Fresh-Check®Life Lines integrator is available as self-adhesive labels, that are attached to the packing material of perishable produce to assure the quality of products to customers. It is based on the principle of color change, which is due to a polymer that has diacetylene monomeric units. It includes a small ring of polymer surrounded by another ring for color reference, the rate of change of color depends on the rate of food quality loss. The color changes from light to dark as the temperature increases.

Vitsab® indicator is based on enzymatic reactions. It has two compartments, one for enzyme plus a dye and the other for substrate (primarily triglycerides). It consists of a bubble-like dot and it is activated by applying pressure, which results in the compartments getting mixed. Because of the reaction between enzyme and substrate, there will be a change in pH and also a change in color. Initially, the dot is in green color and slowly changes to yellow as the product reaches the end of shelf life. The reaction is irreversible and the rate of reaction is directly proportional to

Food/treatment	Packaging materials/methods	Shelf life
Peach, cauliflower, truffle	Tray: PP; Cover: PE-LD/polyethylene terephthalate (PET) (40 μm), 0–14 microperforated package, all wrapped in PE	4 days at 4°C
Strawberry	Stretch PVC	8 days at 1°C
Minimally processed fruits (kiwi, banana and prickly pear)	1. PE/Al/PET	4–12 days at 5°C
	2. Coex. polyolefinic high permeable film	
Sweet cherry	5% O_2 + 10% CO_2	80 days at 1°C
	PE: 13–18% O_2 + 2–4% CO_2	40 days at 1°C
	70% O_2 + 0% CO_2	20 days at 1°C
	Air	30 days at 1°C
Cactus pear fruits	Cryovac MY 15 Plastic box	9 days at 4°C
Carrots, minimally processed	PP + cPP/OPP in:	
	5% O_2/10% CO_2/85% N_2	2 days at 4°C
	80% O_2/10% CO_2/10% N_2	7 days at 4°C
Cabbage, shredded	OPP (30 μm)	9–10 days at 3°C
Cabbage, shredded	Glass jar; PE (30 μm); PP (30 μm) in:	7 days at 0 and 10°C
	Air	
	100% N_2,	
	MAP 1: 100% N_2,	
	MAP 2: 5% O_2/95% N_2,	
	MAP 3:10% O_2/90%	
	MAP 4: 70% O_2/30% N_2 and 100% O_2	
[12].		

Table 5.
Packaging materials and methods effect on the shelf life of fruits and vegetables.

the temperature. Single dot tags are used at consumer level packing for monitoring pallets and cartoons. *ripeSense* is the world's first intelligent ripeness indicator.

The Institute of Food Technologists in the United States has defined shelf life as "The period between the manufacture and the retail purchase of a food product, during which time the product is in a state of satisfactory quality in terms of nutritional value, taste, texture, and appearance". Various factors affecting shelf life are product characteristics, which include intrinsic factors, such as water activity, pH, microflora, availability of oxygen, reduction potential; and extrinsic factors, such as temperature, rainfall, humidity, light, etc., enzymic reactions, chemical reactions, and non-enzymic reactions (**Table 5**).

There are various chemical, biochemical and physical reactions that lead to food quality deterioration. These include enzymic and non-enzymic browning, fat oxidation, hydrolysis, lipolysis, and proteolysis that change the physical and chemical composition of food [33].

5. Conclusion

Recently, the food packaging process, biotechnology, sensor science, information technology, nanotechnology, and other scientific disciplines are coming together to develop a breakthrough in postharvest packaging systems. These improved postharvest handling techniques are continuously getting advanced by creating new opportunities in food industries to utilize technologies in the future Proper and good packing is essential in providing quality products to customers. It is the connecting link between producers and consumers, so it must be done so perfectly to retain the product quality and also customer confidence. The food packaging industry gets highly competitive due to consumer's desire for tasty and slightly processed food products with longer shelf life at a lower cost than their existing packaging. The recent trend in the change of lifestyle leads the food industry well aware of consumer's needs, and therefore, the packaging industry must innovate or stagnate. This condition has posed a great challenge for the food packaging sector to innovate new food packaging techniques. Consumers will often actively seek the freshness of the product with the longest remaining shelf life. Nowadays, novel food packaging technologies, such as active packaging, aseptic packaging, intelligent packaging, nano-packaging, and bioactive packaging intentionally associated with food products have proved to be the best technological research areas. Advances in packaging technology may prevent food spoilage by retarding water penetration, ultraviolet interactions, oxygenation, and ripeness. It is predicted that the future packing material includes radio frequency identification tags. Radio-frequency identification (RFID) tags are advanced forms that can trace and identify a product. Therefore, continuous innovations in active and intelligent packaging systems are expected to secure food quality, safety, and stability and to satisfy the ever-growing need of consumers.

DOI: http://dx.doi.org/10.5772/intechopen.101124

Author details

Trina Adhikary[1]* and Durga Hemanth Kumar[2]

1 College of Horticulture and Forestry, Punjab Agricultural University, Ludhiana, Punjab, India

2 College of Horticulture, Dr. Y.S.R. Horticultural University, Venkataramannagudem, Andhra Pradesh, India

*Address all correspondence to: trinaadhikaryhort@gmail.com

References

[1] Geeson JD. Modified atmosphere packaging of fruits and vegetables. In: International Symposium on Postharvest Handling of Fruits and Vegetables. Leuven, Belgium: Proceeding copy; 1988. pp. 143-147

[2] Moleyar V, Narasimham P. Modified atmosphere packaging of vegetables: An appraisal. Journal of Food Science. 1994;**31**:267-278

[3] Wang CY, Qi L. Modified atmosphere packaging all eviates chilling injury in cucumbers. Postharvest Biology and Technology. 1997;**10**(3):195-200

[4] Tarver T. Food nano technology. Food Technology. 2006;**60**(11):22-26

[5] Smith JP, Ramaswamy HS. Packaging of fruits and vegetables. In: Processing Fruits: Science and Technology. Lancaster, PA: Technomic Publishing Co.; 1996. pp. 379-427

[6] Paine FA, Paine HY. A Handbook of Food Packaging. Leonard Hill; 1983

[7] Bramklev C, Olsson A, Orremo F, Wallin C. Unveiling the concept of packaging logistics. In: Conference Proceeding of NOFOMA. Proceeding copy; 2001

[8] Majid I, Ahmad Nayik G, Mohammad Dar S, Nanda V. Novel food packaging technologies: Innovations and future prospective. Journal of the Saudi Society of Agricultural Sciences. 2016;**17**(4):454-462. DOI: 10.1016/j.jssas.2016.11.003

[9] Thompson AK. Postharvest Technology of Fruit and Vegetables. Oxford: Blackwell; 1996

[10] Chung D, Yam KL. Antimicrobial packaging material containing propyl Paraben. In: IFT Annual Meeting Technical Program: Book of Abstracts. Chicago: Institute of Food Technologists; 1999. p. 21

[11] Day BPF. Recent developments in active packaging. South African, Food & Beverage Manufacturing Review. 1999;**26**(8):21-27

[12] Guilbert S, Gontard N, Gorris LGM. Prolongation of the shelf life of perishable food products using biodegradable films and coatings. Food Science and Technology. 1996;**29**:10-17

[13] Han JH. Antimicrobial food packaging. Food Technology. 2000;**54**(3):56-65

[14] Krochta JM, Baldwin EA, Nisperos Carrido M. Edible Coatings and Films to Improve Food Quality. Lancaster, PA: Technomic Publishing Co; 1994

[15] Lee DS, Kang JS, Renault P. Dynamics of internal atmosphere and humidity in perforated packages of peeled garlic cloves. International Journal of Food Science and Technology. 2000;**35**:455-464

[16] Powrie WD, Skura BJ. Modified atmosphere packaging of fruits and vegetables. In: Ooraikul B, Stiles ME, Horwood E, editors. Modified Atmosphere Packaging of Food. New York, USA; 1991. pp. 169-245

[17] Day BPF. Novel MAP. A brand-new approach. Food Manufacture. 1998;**73**:22-24

[18] Lee DS, Haggar PE, Lee J, Yam KL. Model for fresh produce respiration in modified atmospheres based on principles of enzyme kinetics. Journal of Food Science. 2006;**56**(6): 1580-1585

[19] Alique R, Martinez MA, Alonso J. Influence of the modified atmosphere packaging on shelf life and quality of

Navalinda sweet cherry. European Food Research and Technology. 2003;**217**(5): 416-420

[20] Montanez JC, Rodriguez FAS, Mahajan PV, Frias M. Modelling the effect of gas composition on the gas exchange rate in perforation-mediated modified atmosphere packaging. Journal of Food Engineering. 2010;**96**(3): 348-355

[21] Montanez JC, Rodriguez FAS, Mahajan PV, Frias M. Modelling the gas exchange rate in perforation-mediated modified atmosphere packaging: Effect of the external air movement and tube dimensions. Journal of Food Engineering. 2010;**97**(1):79-86

[22] Silva FM, Chau KV, Brecht JK, Sargent SA. Modified atmosphere packaging for mixed loads of horticultural commodities exposed to two postharvest temperatures. Postharvest Biology and Technology. 1999;**17**(1):1-9

[23] Diab T, Biliaderis C, Gerasopoulos D, Sfakiotakis E. Physicochemical properties and application of pullulan edible films and coatings in fruit preservation. Journal of the Science of Food and Agriculture. 2001;**81**:988-1000

[24] Franssen LR, Krochta JM. Edible coatings containing natural antimicrobials for processed foods. In: Roller S, editor. Natural Antimicrobials for the Minimal Processing of Foods. Boca Raton, Florida: CRC Press; 2003

[25] Suppakul P, Miltz J, Sonneveld K, Bigger SW. Active packaging technologies with an emphasis on antimicrobial packaging and its applications. Journal of Food Science. 2003;**68**:408-420

[26] Berkesch S. Biodegradable Polymers. A Rebirth of Plastic. 2005. pp. 1-14. Available from: https://www.iopp.org/files/public/BerkeschShellieMSUBiodegradablePlastic.pdf [Accessed: April 12, 2010]

[27] Uyama H, Kuwabara M, Tsujimoto T, Nakano M, Usuki A, Kobayashi S. Green nanocomposite from renewable resources: Plant oil-clay hybrid materials. Chemistry of Materials. 2003;**15**:2492-2494

[28] Sorrentino A, Gorrasi G, Vittoria V. Potential perspectives of bio-nanocomposites for food packaging applications. Trends in Food Science and Technology. 2007;**18**:84-95

[29] Bharadwaj. Modeling the barrier properties of polymer layered silicate nano omposites. Macromolecules. 2001;**34**(36):9189-9182

[30] Sorrentino A, Gorrasi G, Tortora M, Vittoria V. Barrier properties of polymer/clay nanocomposites. In: Mai Y-W, Yu Z-Z, editors. Polymer Nanocomposites. Cambridge, UK: Wood-head Publishing Ltd.; 2006. pp. 273-292

[31] Rhim JW, Lee JH, Kwak HS. Mechanical and barrier properties of soy protein and clay mineral composite films. Food Science and Biotechnology. 2006;**14**:112-116

[32] Xu Y, Ren X, Hanna MA. Chitosan/clay nano-composite film preparation and characterization. Journal of Applied Polymer Science. 2006;**99**:1684-1691

[33] Robertson GL. Food Packaging-Principles and Practice. 3rd ed. Boca Raton: CRC Taylor and Francis Group; 2012

Chapter 8

Postharvest Technology of Tamarind

P. Sudha, P. Rajkumar, A. Astina Joice, I.P. Sudagar and R. Arulmari

Abstract

Tamarind is a multi-purpose long-lived tree with heavy drooping branches and thick foliage. The entire fruit consists of 55% pulp, 34% seeds, and 11% hull and fibers. The tamarind tree produces numerous elongated fruit pods in a season that encompasses its branches in myriad. Brittleness in shell, changes in testa color, and a hollow sound from fruit when finger pressed signify matured fruit of the tree. Postharvest operations involved in Tamarind are drying, dehulling, defining, deseeding, pressing into cake, and storage. These operations are carried out by traditional and mechanical methods. Tamarind dehullers and deseeder were developed with efficiencies of around 94% and 83% respectively to minimize the losses involved in manual handling. The intrinsic value of raw tamarind may be furthermore desirable through processing into value-added products.

Keywords: tamarind, drought conditions, fruit pods, harvesting, intrinsic value, processing

1. Introduction

Tamarind (*Tamarindus indica* L.) is a commercially important tree that can be found in many Asian, African, and South American nations. The tree can reach its full potential with a crown diameter of 12 m and a height of 25 m. It is perfect for dry and semi-arid climates, particularly in drought-prone locations that lasts for a long time. The tamarind tree is a low-maintenance tree that is easy to grow. It is largely devoid of major pests and illnesses, and it has a long life expectancy. It can live up to 80–200 years and produce 150–500 kg of pods [1]. Each pod has a firm outer shell that surrounds a deep brown mushy pulp that encases two to ten hard dark-brown seeds. The pulp is sticky, as it is highly hygroscopic. The tamarind pulp is high in sugar, ranging from 21% to 30%, and its hygroscopicity increases as the relative humidity rises at room temperature.

India is the world's greatest producer of tamarind, with 300,000 t projected to be produced each year. It's especially common in states like Madhya Pradesh, Bihar, and Andhra Pradesh, Karnataka, Tamil Nadu, and West Bengal. Thailand is the second greatest producer, with 150,000 t recorded in 1995, with the sweet variety accounting for the majority of tamarind [2]. Mexico is also on the list commercially produces tamarind to a volume of approximately 29,600 t per year (**Figure 1**).

Figure 1.
Tamarind tree, India.

2. Postharvest handling practices

2.1 Harvest operations

The yield of pods stabilizes about 15 years and can last up to 50 or 60 years. When tamarind fruits are finger pressed, a hollow and loose sound is generated, indicating that the pulp has shrunk with maturity and the fruit is ready for harvesting. Brittleness develops in the shell. Additionally, a change in testa color could signify matured fruit of the tree [3]. Individual fruits on the same tree mature at different periods, making it difficult to select the right one for which harvesting is required. Pods are picked at various stages of ripeness, depending on their intended purpose. The sour tamarind mature fruits are commonly sold in most nations. It can be obtained by shaking trees and gathering up fallen fruits. The fruits are usually allowed to ripe on the tree before being harvested, reducing the moisture content to around 20% [4]. If left unharvested, the pods can stay on the tree for over a year after flowering, and will inevitably descend. Fruits for immediate processing are frequently harvested by separating the pod from the stalk, leaving longitudinal fibers connected. Fungi and beetles in humid conditions, ripe fruit is more easily attacked, thus they should be harvested before they have reached full ripeness.

Trunk shakers and branch shakers are the few mechanical advancements in harvesting fruit bunches. Trunk shakers are best suited for trees that have soft trunks and branch shakers are best suited for trees that have hard trunks like tamarind. Nowadays, branch shakers are available at a 2.5 kW power level. A two-stroke petrol engine drives the shaker. This commercial branch shaker is similar to a brush cutter with a 2 m long hooked pipe at the end to hold the branches intact and the reciprocal action of the engine imitates the shaking action [5].

3. Engineering properties of tamarind

The understanding of physical and mechanical qualities aids in the analysis of its behavior during handling and the design of process equipment. The average value of properties such as moisture content, size, shape, bulk density, true density, porosity, angle of repose, surface area, and coefficient of friction was determined for whole and dehulled tamarind (**Table 1**) [6].

4. The traditional method of tamarind processing

In general, tamarind processing is done using both wet and dry processes. Tamarind is usually processed in a dry manner, with the process of drying (the whole tamarind to avoid pulp sticking to the hull), dehulling, defining, deseeding, pressing into cakes, and storage being the most prevalent process. The dehulling of tamarind entails sun drying the harvested entire tamarind and then pounding the hull away from the pulp with sticks (**Figure 2**).

It is necessary to remove the seeds and it's one of the most essential and labor-intensive processes (**Figure 3**). In the northern region of Tamil Nadu, the current method of deseeding tamarinds is by the manual pounding of the vertically aligned fruits with a hammer or wooden mallet by female laborers (**Figure 4**). Tamarind is deseeded by hand pounding, where a stone mortar is coated with oil, usually castor oil, and a wooden pestle is used to exert impact stress. A knife or a long sharp needle is also used to remove the seeds. The traditional approaches are rough, unhygienic, labor intensive, and time-consuming. As a result, to make tamarind processing easier, state agricultural universities created and released machinery for the benefit of processors and to ensure hygienic practices in processing (**Figure 5**).

Engineering properties	Whole tamarind	Dehulled tamarind
Geometric mean diameter	33.87 mm	26.21 mm
Sphericity	0.33	0.28
Surface area	3000.18 mm^2	1904.98 mm^2
Bulk density	240.39 kg/m^3	612.24 kg/m^3
True density	469.59 kg/m^3	1182.41 kg/m^3
Porosity	48.80%	48.22%
Angle of repose	35.4°	39.4°

Table 1.
Engineering properties of whole and dehulled tamarind fruit.

Figure 2.
Labors beating the dried whole tamarind as part of dehulling.

Figure 3.
Women workers are deseeding tamarind.

Figure 4.
Defibering action.

Figure 5.
Pressing action of deseeded tamarind into rubber ring to form cake.

5. Drying of whole and dehulled tamarind

Dehulling and deseeding are the significant tasks in tamarind processing for which tamarind is exposed to drying to moderate the tenacity with physical and mechanical parts. A local variety of tamarind dried under the sun as shown (**Figure 6**) followed by plate drying at temperatures of 50, 60, and 70°C. Mechanical drying of tamarind at 70°C had a higher dampness expulsion rate followed by drying at 60°C and 50°C. The drying data for sun drying of tamarind was well fitted with Midilli et al. model.

6. Mechanization

6.1 Tamarind dehuller

A dehuller for tamarind was developed at Tamil Nadu Agricultural University (TNAU), Coimbatore, India (**Figures** 7 and **8**). It has a capacity of 100 kg/h with dehulling productivity of 94%. The impact force and sieve shaking mechanism were used to develop the Tamarind dehuller. The impact force from the rotating beaters was applied to the dried tamarind fruit fed through the feed hopper. The outlet received the dehulled, un-hulled, and hulled fruits. The system includes sharp "L" shaped pegs made up of 15 × 3 mm size gentle steel level mounted on the focal shaft that encased with 20 × 5 mm mild steel oblong sieve [7].

6.2 Tamarind deseeder

A tamarind deseeder developed at Tamil Nadu Agricultural University with the principle of impact and simultaneous shear (**Figure 9**) is used widely in Krishnagiri and Dharmapuri districts of Tamil Nadu to deseed small-sized tamarinds. The deseeding efficiency of the machine is 83% and the cost is Rs. 20,000/– [6].

A hammer-type tamarind deseeder was created and evaluated at Kumulur, Tamil Nadu Agricultural University comprises a feeding roller with rubber lining, hammering mechanism, motor, and power transmission framework and casing to

Figure 6.
Sun drying of the whole tamarind.

Figure 7.
Tamarind dehuller developed at Tamil Nadu Agricultural University, Coimbatore.

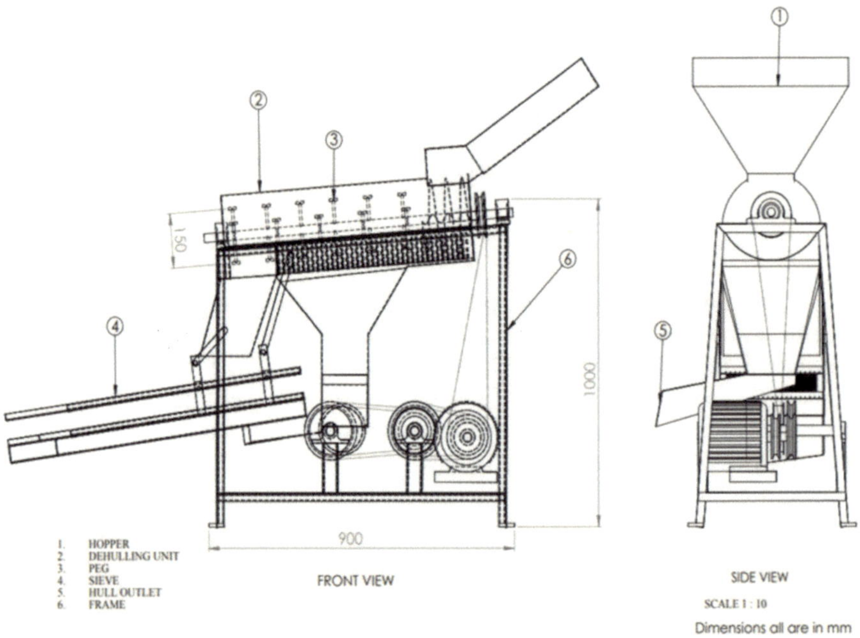

Figure 8.
Tamarind dehuller developed at Tamil Nadu Agricultural University, Coimbatore.

help the mechanism. The machine work on the standard of impact and deseeding proficiency of the machine is discovered to be 79% at 5 rpm (0.06 m/s); the peripheral speed of feeding roller to deseed the tamarind at the moisture content of 22.5% on a dry basis gives minimal mechanical harm to seed (0.3%) and pulp (14.94%). The equipment suits the large-sized fruits with multiple seeds. Feeding of small-sized fruits with single and two-seeded fruits in the feeding roller is troublesome due to the variation in shape.

6.3 Hammering mechanism

Link mechanism was used to imitate hammering action over tamarind fruit. 400 mm long and 25.4 mm diameter polished rod was mounted with lateral frame utilizing 25.4 mm diameter bearing block over the lateral frame. One end of the

Figure 9.
Tamarind deseeder developed at Tamil Nadu Agricultural University, Coimbatore.

Figure 10.
A hammer-type tamarind deseeder is available at Tamil Nadu Agricultural University.

polished rod was fitted with FPS 100 pump bearing which touches the round flange to transmit reciprocating motion. A mild steel flat of 40 × 12 mm was spring-loaded and mounted on the shaft within the space between the bearing block to imitate hammering action and the height of the flat was 300 mm. The top end of the flat was screwed with a wooden portion to imitate a wooden hammer (**Figure 10**) [8].

7. Value addition and by-products

7.1 Tamarind pulp

The most important and generally utilized part of the tamarind tree is its natural product pulp. It establishes 30–50% of the ripe fruit, the shell and fiber represent 11–30%, and the seed around 25–40% [9]. Pulp is rich in pectin and reducing sugars and contains critical measures of organic acids, 98% of which is tartaric acid. It is a unique plant acid that is generated from the principal carbohydrate products of photosynthesis and is not utilized metabolically by the plant once formed. The primary flavor compound of the pulp is 2-acetyl furan. The quantity of tartaric acid does not diminish as the fruit ripens, implying that it is stable, in the development of fruit. Reducing sugars grow to 30–40% during this stage of fruit development, giving the sour fruit a sweeter taste.

Tamarind pulp was assessed based on its physicochemical properties such as crude protein, crude fiber, fat, ash, moisture content, water activity (Aw), particle shape, particle size distribution, and density (**Figure 11** and **Table 2**) [10].

Figure 11.
Tamarind fruit pulp.

S. no.	Proximate analysis (%)	Tamarind pulp +10% w/v maltodextrin
1.	Moisture content	5.15 ± 0.15
2.	Crude protein	0.43 ± 0.02
3.	Crude fiber	79.92 ± 0.85
4.	Ash	17.80 ± 2.59
5.	Fat	0.43 ± 0.18
6.	Water activity (A_w)	0.69 ± 0.01

Table 2.
Proximate analysis of tamarind pulp.

8. Processed products from tamarind pulp

8.1 Tamarind beverage

The tamarind fruit pulp is used for the preparation of beverages. It can be used to make high-quality ready-to-serve beverages, syrups, and concentrates with a six-month shelf life when stored at room temperature [11]. On a small scale, the fruit pulp is dissolved in water and squeezed by hand to make a delightful drink. Water is added to dilute the drink after the extraneous substance is removed. Tamarind pulp is a delightful drink in southern and central America, as well as Asia. Arnold [12]

Component	F1 (g)	F2 (g)	F3 (g)	F4 (g)
Tamarind pulp	100	120	80	80
Sugar	40	30	30	40
Purified water (mL)	500	500	500	500

Table 3.
Design of formulations to prepare the natural tamarind beverage.

Parameter	F1	F2	F3	F4
pH	2.73 ± 0.08	2.88 ± 0.02	2.84 ± 0.01	2.83 ± 0.01
TSS (Bx)	12.36 ± 0.39	10.83 ± 0.06	9.70 ± 0.10	10.73 ± 0.15

Table 4.
Results of pH and total solids of the natural beverage of tamarind before pasteurizing.

studied the physicochemical properties of natural tamarind beverages under four different formulations (**Tables 3** and **4**).

8.2 Tamarind syrup

Tamarind syrup is created by softening immature fruit pulp and straining it through a cheesecloth. The pulping process involves breaking the shells by hand and agitating the pulp and seeds in water to separate the pulp and seeds. A half-teaspoon of baking soda is poured into a cup of juice. The mixture is reduced to one-half of its original volume while also eliminating the scum that has risen to the surface. The juice is strained once more. A quarter cup of sugar is added to every cup collected. After 20 min, the mixture is boiled once more. The cooled syrup is poured into the container bottles that have been sterilized and sealed. The suggested tamarind pulp content in syrup is 20–24% to make a beverage with distinct flavor and acidity. This syrup comprises 56.7% total solids and 43.8% reducing sugar tartaric acid a total acidity of 1.11% as tartaric acid.

9. Tamarind juice concentrate

Tamarind concentrate is an acid pulp concentrate made from tamarind pulp that is free of skin, seed, fibers, and other impurities. The pulp from the tamarind pods is collected, and the juice is extracted from the pulp. The shells are broken by hand and agitated in water during the pulping process. The suggested tamarind pulp content in syrup is 20–24% to make a beverage with a distinct flavor and acidity [13]. By adding gelatine to clarified tamarind juice, the structural and colloidal phases are fully eliminated. The juice is translucent and retains its color and flavor. Under vacuum, insoluble particles are removed and soluble extracts are concentrated (**Figure 12**).

Deepika [14] studied Spray Drying of Tamarind Juice Concentrate and Powder Characteristics, the results are shown in **Table 5**.

9.1 Tamarind pulp powder

Tamarind pulp powder (TPP) is a convenience food manufactured from tamarind pulp that has been concentrated, dried, and milled into a powder. TPP is made

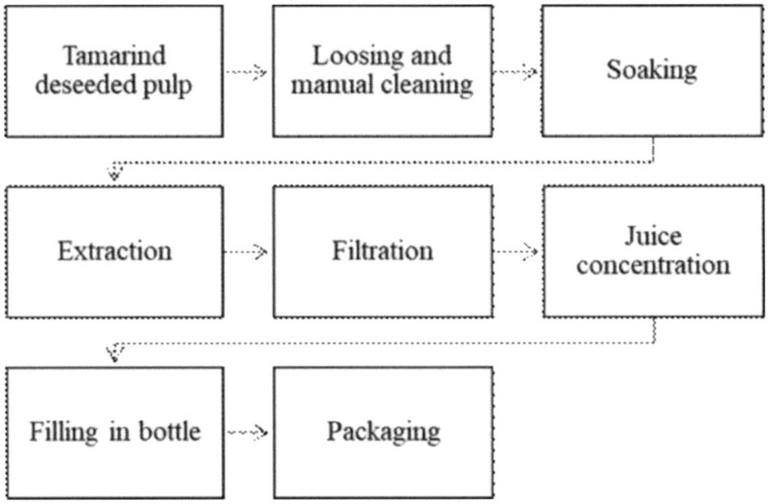

Figure 12.
Processing of tamarind juice concentrate.

Parameters		Mean value
Moisture content (wb)		30.13
Water activity, a_w		0.836
pH		1.63
Bulk density (g/cm^3)		0.905
Color	L	2.04
	a	1.78
	b	1.77
Tartaric acid (%)		13.07
Total sugars (%)		41.2
Protein (%)		2.1
Crude fiber (%)		2.0

Table 5.
Physicochemical properties of tamarind juice concentrate.

by concentrating, drying, and grinding tamarind pulp into a fine powder. In many Indian recipes and sauces, tamarind powder is used as a condiment/adjunct and souring ingredient. It has a large market potential as a convenient product.

9.2 Foam mat dried tamarind pulp powder

Liquid foods are whipped into stable foams and then air-dried using the foam mat drying process. Lower drying temperatures and faster drying times are the key benefits of foam-mat drying processes. The larger surface area exposed to the drying air, which speeds up the drying process, is responsible for these advantages.

The lack of foam stability throughout the heating cycle, however, should be taken into account. Cellular breakdown occurs when the foam does not remain stable, presenting major problems during the drying process. This drawback can be overcome by using a film stabilizer, such as polymeric material. Vernon-Carter et al.

[15] reported drying tamarind pulp using foam mat drying with various foaming agents. Mesquite gum, ovalbumin, and a low molecular weight surface active blend were hydrated to 50% (w/w) solutions and applied to the samples single or in combination (**Figure 13**).

9.3 Spray-dried tamarind juice concentrate (TJC) powder

See **Figure 14** and **Table 6**.

9.4 Drum drying of tamarind pulp

Weerachet et al. [16] studied the production of tamarind powder by drum dryer using Maltodextrin (MD) and Arabic gum (AG) as adjuncts (**Figures 15** and **16** and **Table 7**).

Figure 13.
Tamarind pulp powder.

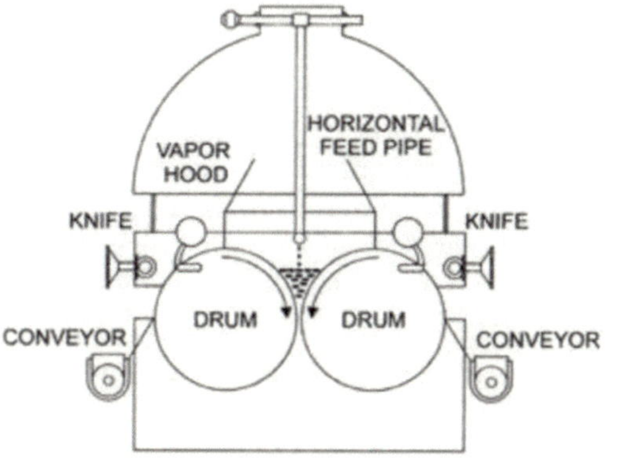

Figure 14.
Flow chart for the preparation of spray-dried TJC powder.

S. no	Quality attributes	Spray-dried TJC powder			
		Storage period (days)			
		1	30	60	90
1.	Moisture content (w.b)	2.14	2.86	3.29	3.89
2.	Water activity, a_w	0.263	0.298	0.329	0.399
3.	Bulk density, g/cm^3	0.492	0.522	0.586	0.602
4.	Tartaric acid (%)	9.87	10.84	11.85	13.91
5.	Solubility (%)	88.3	83.21	78.75	72.83
6.	Wettability (s)	76	91	125	154
7.	Dispersibility (%)	80.16	74.65	69.00	62.34
8.	Color value				
	Redness ('a' value)	15.84	13.20	10.89	7.99
	Yellowness ('b' value)	24.07	22.15	21.32	19.56

Table 6.
Quality changes in spray-dried TJC powder during storage packed in Aluminum foil pouches [14].

Figure 15.
Schematic diagram of drum dryer.

9.5 Fruit leather

Fruit leather prepared from the dried sheets of tamarind fruit pulp will have a soft, rubbery texture and a sweet taste (**Figure 17**) [17]. Ghada et al. [18] studied the effect of different drying methods (cabinet drier (70°C) and solar drier (54 ± 4°C)) on the quality and consumer acceptability of tamarind leathers (**Figure 18**). Results showed that drying methods influence the color changes of tamarind leather. Effect of drying on quality characteristics are shown in **Table 8.**

9.6 Tamarind candy

Tamarind candy is one of the most liked products by consumers because of its natural sour-sweet blend. Candies are prepared after boiling tamarind pulp with a sufficient amount of sugar and cooking it with a very little amount of water. Arghya

Figure 16.
The processing flow chart for drum drying of tamarind pulp.

Drying condition				Bulk density (g/ml)	Moisture content (% wb)	Water activity	Solubility (second)
No.	Drying aid	Drying temperature (°C)	Ratio of tamarind juice and drying aid				
1	MD	120	1:1.4	$0.816^{a} \pm 0.066$	$3.46^{a} \pm 0.03$	$0.260^{a} \pm 0.003$	$83^{a,b} \pm 2$
2	MD	140	1:0.8	$0.781^{a} \pm 0.045$	$3.38^{b} \pm 0.03$	$0.326^{b} \pm 0.006$	$98^{a} \pm 18$
3	MD	140	1:1.4	$0.478^{b} \pm 0.063$	$3.11^{c} \pm 0.04$	$0.342^{c} \pm 0.001$	$8^{c} \pm 1$
4	AG	120	1:0.4	$0.731^{a,c} \pm 0.038$	$3.20^{d} \pm 0.04$	$0.306^{d} \pm 0.007$	$79^{a,b} \pm 18$
5	AG	120	1:0.8	$0.790^{a} \pm 0.037$	$3.62^{c} \pm 0.04$	$0.265^{a,c} \pm 0.006$	$73^{b} \pm 8$
6	AG	140	1:0.4	$0.648^{c} \pm 0.032$	$3.09^{c} \pm 0.02$	$0.276^{c} \pm 0.002$	$140^{d} \pm 8$
7	AG	140	1:0.8	$0.783^{a} \pm 0.094$	$3.41^{a,b} \pm 0.03$	$0.263^{a} \pm 0.012$	$16^{c} \pm 2$

Note: Bulk density, color, moisture content, water activity and solubility values are mean ± standard deviation (n = 3). Means with the same superscript within same column are insignificant different (P > 0.05).

Table 7.
Bulk density, moisture content, water activity, and solubility of tamarind powders.

Figure 17.
Process flow chart for developing tamarind fruit leather.

Figure 18.
Tamarind fruit leather.

Parameter	Cabinet drier	Solar drier	S.E	LSD (5%)
Texture	3.29 ± 0.31	2.52 ± 0.36	0.336	0.762
Color	0.138 ± 0.01	0.043 ± 0.03	0.010	0.022
Rehydration ratio	1.44 ± 0.16	1.78 ± 0.26	0.214	0.484
Drying ratio	3.50 ± 0.00	3.25 ± 0.06	0.039	0.088
pH-value	2.78 ± 0.03	2.81 ± 0.03	0.029	0.067
Titratable acidity	6.86 ± 0.03	7.83 ± 0.39	0.274	0.622

Table 8.
Effect of different drying methods on quality characteristics of tamarind leathers.

Mani et al. [19] standardized the recipe for the preparation of Tamarind candy (**Figures 19** and **20**).

9.7 Tamarind pickle

To make tamarind pickle, the commercially available pulp is used. Pickles are hot, spicy, and salty-sour in flavor, and they can be stored for months. The inclusion of salt, enhanced acidity, and spices aid in preservation (**Figures 21** and **22**).

Figure 19.
Flow chart for tamarind candy preparation.

Figure 20.
Tamarind candy.

Figure 21.
Flow chart for tamarind pickle preparation.

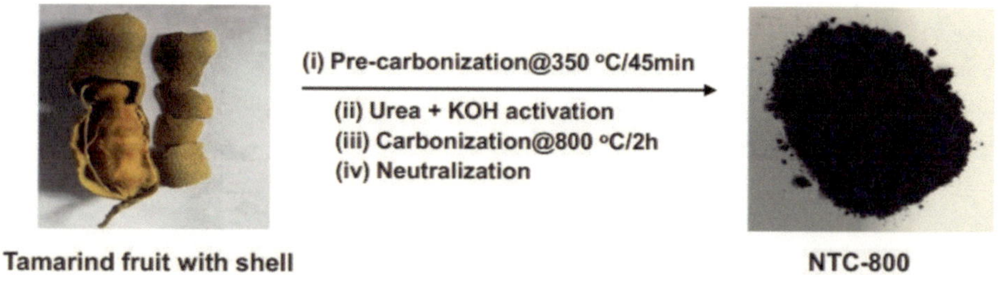

Figure 22.
Tamarind pickle.

9.8 Tamarind ketchup

Clean the tamarind pulp, then boil it in freshwater to extract the tamarind puree. Cook on medium heat with 10% sugar and 1% salt. Then take it off the fire and combine it with the spices. Boil the edible oil in a saucepan and put sliced ginger, small bits of garlic, and chili along with the product and cool down the product before packing (**Figure 23**).

9.9 Champoy

Tamarind fruits can also be made into balls, or "champoy," a popular tamarind snack in the Philippines. Two cups of boiled and mashed sweet potato, two cups of

Figure 23.
Tamarind ketchup.

sugar, a one-eighth cup of salt, and one cup of water are added to one cup of pulp with seeds. The mixture is heated over low heat, stirring constantly, until it thickens and can be molded into balls (**Figures 24** and 25).

9.10 Tamarind ade

This is a delicious tamarind drink made in the Philippines and several tropical American countries by blending ripe pulp with sugar and water until it reaches the desired taste. Making Ade at home is as simple as shelling the fruits, placing them in 2–3 L of water, allowing it to stand for a brief time, then adding a tablespoon of sugar and vigorously shaking Spices like cloves, cinnamon, ginger, pepper, or lime slices are sometimes added to increase the flavor [20].

Figure 24.
Tamarind balls.

Figure 25.
Tamarind Ade.

10. By-products of tamarind

10.1 Tamarind seed oil

Tamarind seed kernels produce an amber-colored oil that is odorless and sweet in flavor, similar to linseed oil. It is used in varnishes, paints, and lamp oil [2, 20], but it is also considered to be pleasant and of culinary quality [2, 20]. Tamarind oil has an iodine content of less than 100 mg/100 g, making it a non-drying oil (**Figure 26**).

10.2 Fruit shell

Tamarind shell is used as the carbon precursor for generating the activated carbon (AC) and the resultant AC materials utilized for water purification and supercapacitor applications. Vengatesan et al. [21] studied the tamarind shell-derived N-doped carbon for capacitive deionization (CDI) (**Figure 27**).

N-doping is proposed to be an effective method in not only improving the electrical conductivity and wettability of the carbon but also played a crucial role in

Figure 26.
Tamarind seed oil.

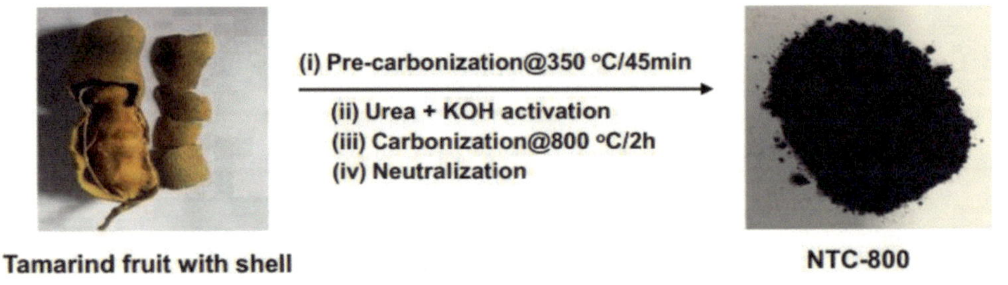

Figure 27.
Schematic representation of the formation of NTC-800.

enhancing the electro-sorption performance. As such, the low-cost biomass waste tamarind shell derived N-doped carbon nanosheets developed offer a promising electrode material for conventional high-performance symmetric CDI applications.

10.3 Bark

Tamarind barks and leaves contain a yellowish or brownish bitter-tasting organic substance called Tannin. The bark has 70% of tannin and found a great place for its usage in the tanning industry. Bark tannins are utilized in the production of ink and the fixation of colors in Zambia [2]. Many other African countries use the bark to manufacture ink. Tamarind twigs are used as "chewsticks," while the bark is utilized as a "chewing gum" masticatory, either alone or as a lime replacement in betel nut [2]. Hordenine is an alkaloid found in the bark [20].

Lac: Tamarind tree is a host for lac insect that deposits a resin on the twigs. This product should be harvested and sold as a stick lac, but it is not considered an important source.

10.4 Tamarind seed gum

Tamarind gum is obtained from the endosperm of seeds of the tamarind tree, which is a seed gum with potential industrial applications [22]. Tamarind gum is having applications in paper, food, textile industry, etc. The composition of tamarind kernel, the source of gum, resembles the cereals. With 15.4% to 12.7% protein, 3–7.5% oil, 7–8.2% crude fiber, 61–72.2% non-fiber carbohydrates, 2.45–3.3% ash (d.b) [23].

11. Conclusion

Tamarind is a crop that can be eaten as a fruit or used as a condiment. The fruit has a delectable sour-sweet flavor. It's full of vitamins and minerals, as well as antioxidants. Due to increased knowledge of Good Manufacturing Practices (GMP) and labor scarcity, mechanized tamarind processing without human intervention is expected [24]. To make tamarind processing easier, state agricultural universities developed and distributed machinery for the advantage of processors and to assure hygienic processing techniques. Tamarind can be used to produce many values added and by-products so that it fetches more market price to the producers Physicochemical, thermal, structural changes take place during value addition of tamarind. Research in these areas is carried out to optimize changes in properties.

Author details

P. Sudha*, P. Rajkumar, A. Astina Joice, I.P. Sudagar and R. Arulmari
Department of Food Processing Engineering, Agricultural Engineering College and Research Institute, Coimbatore, Tamil Nadu, India

*Address all correspondence to: sudhatnau@gmail.com

References

[1] Mathew KM. In: Peter KV, editor. Handbook of Herbs & Spices. England: Wood Head Publishing Limited; 2001. pp. 512-531

[2] El-Siddig et al. Tamarind (*Tamarindus indica* L.): A review on a multipurpose tree with promising future in Sudan. Journal of Applied Botany—Angewandte Botanik. 2006;**73**: 202-205

[3] Yadev SK. Antimitotic and cytological activities of tropical forests. *Tamarindus indica*. Journal of Tropical Forestry. 2011;**2**(1):53-58

[4] Benero JR et al. Tamarind. Journal of the Agricultural University, Puerto Rico. 1972;**56**(2):185-186

[5] Anonymous. Web Source. 2015. Available from: http://www.aliexpress.com/item/Olive-harvestermachine-nut-olive-shaker-tree-shaking-machine-tree-branch-shakervibration-machine-fruit-branch/32319339259.html

[6] Pandian NKS, Rajkumar P. Development and evaluation of tamarind seed remover [unpublished M.Tech. thesis]. Coimbatore, India: Department of F&AP Engg. TNAU; 2010

[7] Janshi SRJ. Design and development of a power-operated tamarind huller cum deseeder [unpublished M.Tech. thesis]. TNAU Coimbatore: Department of F&AP Engg; 2012

[8] Pandian NKS, Rajkumar P. Development and evaluation of hammer type tamarind (*Tamarindus indica* L.) deseeder. Research Journal of Agricultural Sciences. 2014;**5**(6): 1228-1231

[9] Purseglove JW. Tropical Crops. Dicotyledons. Longman Science and Technology; 1987. pp. 204-206

[10] Taufiq. Physicochemical properties of tamarind and pineapple fruit pulps and powder. International Food Research Journal. 2015;**22**(2):707-712

[11] Kakade. Studies on storage of tamarind and processing of value-added tamarind products [M.Tech thesis]. Krishikosh; 2004

[12] Arnold D et al. Microwave pasteurization for natural tamarind beverage. The Canadian Society for Bioengineering Paper No. CSBE17109. 2017

[13] Bueso CE. Soursop tamarind and chironka. In: Nagy S, Shaw PE, editors. Tropical and Subtropical Fruits. Vol. 1980. Westport, Conn: AVI Publishing; 1980. p. 375

[14] Deepika K. Studies on spray drying of tamarind juice concentrate and powder characteristics [M.Tech. thesis]. Coimbatore: Department of Food and Agricultural Process Engineering, Tamil Nadu Agricultural University; 2008

[15] Vernon-Carter et al. Effect of foaming agents on the stability, rheological properties, drying kinetics and flavour retention of tamarind foam-mats. Food Research International. 2001;**34**(7):587-598

[16] Weerachet et al. Production of tamarind powder by drum dryer using maltodextrin and Arabic gum as adjuncts. Journal of Science & Technology. 2011;**33**(1):33-41

[17] Raab C, Oehler N. Making dried fruit leather. Fact Sheet 232. Oregon State University Extension Service; 1999. pp. 1-4

[18] Abdel Rahman GH et al. Studies on the effect of different drying methods on quality and consumer acceptability

of tamarind leathers. Journal of Agri-Food and Applied Sciences. 2017;**5**(1): 1-5, 28

[19] Mani A et al. Recipe standardization for preparation of Tamarind candy. The Pharma Innovation Journal. 2020;**9**(5): 166-170

[20] Morton J. Fruits of Warm Climates. Miami FL: 1987. pp. 115-121

[21] Vengatesan et al. Tamarind shell derived N-doped carbon for capacitive deionization (CDI). Journal of Electroanalytical Chemistry. 2019; **848**(1):113307. Available from: https://www.ort.purdue.edu/newcrop/morton/tamarind.html

[22] Abo-Shosha et al. Preparation and characterization of polyacrylic acid/karaya gum and polyacrylic acid/tamarind seed gum adducts and utilization in textile printing. Carbohydrate Polymers. 2008;**74**(2): 241-249

[23] Sachinkumar. Tamarind gum: A pharmaceutical overview. Pharmaceutical Reviews. 2008;**6**(4)

[24] Patil SJ, Nadagouder BS. Industrial products from *Tamarindus indica*. In: Proc. Nat. Sym. on *Tamarindus indica* L, Tirupathi (A.P.), Organized by Forest Dept. of A.P., India; 27–28 June 1997. 1997. pp. 151-5

Chapter 9

Processing of Tree Nuts

Chang Chen and Zhongli Pan

Abstract

Tree nuts are consumed as healthy snacks worldwide and are important economic crops. In this chapter, post-harvest processing technologies of tree nuts are discussed, with focus on the drying, disinfection, disinfestation, and downstream processing technologies (blanching, kernel peeling and roasting) for the control and preservation of product quality and safety. Almonds, walnuts, and pistachios are selected as the representative crops for the discussion. Current status, recent advances, and challenges in the scientific research, as well as in the industrial productions are summarized. Some new perspectives and applications of tree nut processing waste and byproducts (such as shells and hulls) are also introduced. The contents presented in this chapter will help both scientists and stakeholders to better understand the tree nut processing and provide technological recommendations to improve the throughput, efficiency, and sustainability of the processes, and preserve the quality and safety of the products.

Keywords: tree nuts, drying, disinfection, disinfestation, food safety, food quality, energy, sustainability

1. Introduction

Tree nuts have high nutrient contents, including oils, proteins, and carbohydrates [1]. Due to their high pleasant flavor and various benefits for human health, tree nuts have gained increasing popularities worldwide, and are consumed as healthy snacks or food ingredients for cooking [2]. The global market of tree nuts was reported at US$88.8 billion in 2020, and it is expected to grow continuously to US$103 billion by 2027 [3]. Commonly consumed tree nuts include almonds, walnuts, pistachios, pecans, macadamia nuts, hazelnuts, and cashews, etc. Among them, almonds, walnuts, and pistachios are the most popular types, accounting for almost 70% of total tree nuts production in the world [4]. The global production mass of almonds, walnuts, and pistachios in 2020 were: 1,700,000, 2,300,000 and 985,000 metric tons, respectively, which increased 15%, 20% and 37%, respectively, compared to the 2019 harvest season [5].

Tree nuts are usually harvested in a relative short harvest season (about 1–2 months from late summer to early fall). The harvested nuts need to fulfill the year-round consumption. Almost all tree nuts are composed of a thick and wet hull that wraps the shell and kernel inside at harvest (**Figure 1**). As the result, freshly harvested tree nuts usually have high initial moisture contents (IMCs) and water activity. Such characteristics make fresh tree nuts vulnerable to spoilage and quality deterioration after harvesting [7, 8]. Therefore, artificial drying is critical to preserve the quality and safety of the nuts. Meanwhile, since tree nuts are rich in unsaturated fatty acids [9, 10], their oil quality is sensitive to the thermal drying

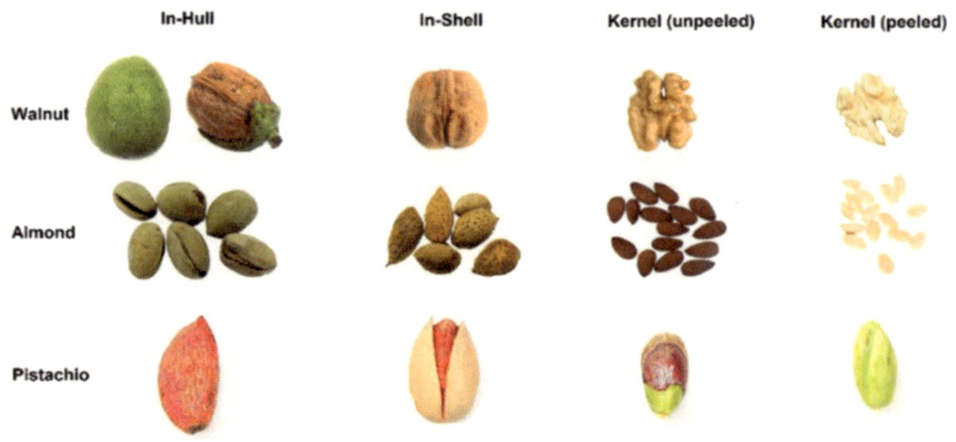

Figure 1.
The photos of walnut, almond and pistachio at different stages of drying (in-hull, in-shell, unpeeled kernel, and peeled kernel) [6].

process [11–13]. Thus, tree nuts need to be dried appropriately and efficiently after harvesting to ensure the quality, safety, and market value of dried products [14–18].

Prior to harvest, plants usually have very good natural defense mechanisms against microbial spoilage. After harvesting, the high moisture contents (MCs) and nutrient contents make them vulnerable to microbial spoilages. Any food safety problems associated with the regional tree nuts production could cause international outbreaks and significantly impact the human health [19–21]. Drying alone usually cannot achieve adequate disinfection and disinfestation. Therefore, further disinfection and disinfestation are critical for extending the shelf life and safety of the dried products [22, 23]. Depending on the type of the final products, further processing, such as roasting, blanching and kernel peeling, may also be needed to produce desired products for consumption.

In addition, thermal and chemical processing of tree nuts are energy intensive and cause significant environmental impacts [24, 25]. It is worthy of noticing that the food production sector is responsible for one-quarter of the world's total greenhouse gas emissions [26] and consumed 200 EJ energy per year [27]. The fast-growing population, increasing production volume and market demands for food production will put more pressure and challenges on the industries for higher processing throughput and efficiency [28]. In the 2021 United Nations Climate Change Conference (COP 26), the world leading countries have committed to achieve 'net zero' carbon neutrality goals by 2050 [29, 30], which needs contributions from all sectors, including the postharvest processing of tree nuts.

In the following sections, the current status, recent advances, and challenges in the postharvest processing technologies are summarized using walnuts, almonds and pistachios as examples.

2. Conventional harvest and postharvest processing methods

Due to similar hull-shell-kernel multiplayer structures (**Figure 1**), the postharvest processing operations, including cleaning, dehulling and drying, etc. are similar for different types of tree nuts. Meanwhile, due to the differences in the shell conditions, MCs and lipid compositions, the processes have some differences.

In California and Australia, almonds are shaken off trees when they are mature and dried on-ground in orchards (**Figure 2**), taking advantage of the hot and dry

Figure 2.
Harvesting and post-harvest processing of almonds: (A) almond shaking; (B) on-ground drying; (C) stockpiling; (D) off-ground harvester (figures obtained from industrial partner or taken by the author's lab).

weather during the harvest season [31, 32]. Conventional drying process takes 7–14 days depending on the weather condition, until the overall MC of whole almonds achieved about 12% on wet basis (while the kernel moisture reaches about 6%). Dried almonds are swept together into windrows and picked up by machineries, which are then stockpiled outdoor for temporary storage. The stockpiles are aerated, which allows the product moisture to equilibrate before mechanically cleaned and de-hulled with abrasive huller [33]. Currently, two major problems with this procedure are: (1) the sweeping and picking up generate large amount of dust, which spreads away in the air and causes pollution [34]; (2) almonds contact the soil directly while drying on-ground, which induces severe insect damage and microbial spoilage [35, 36]. In Europe (mainly Spain), almonds are harvested off-ground and de-hulled in-field, which are then dried in silo dryers with air [31]. Recently, the Almond Board of California (ABC) and Almond Board of Australia are also supporting research in developing off-ground harvesting technology to mitigate the problems of conventional harvesting [37]. Meanwhile, it brings up critical needs to dry the high-moisture almonds artificially and efficiently and to handle the large volume of production in the short harvest season for product quality and safety.

The harvesting method of walnuts are similar to almonds. However, since walnut shell is harder and thicker than that of the almond, natural drying is not efficient enough to dry the inside of them. If the walnuts stay too long on the ground, microbial spoilage becomes significant. Therefore, after drying on the orchard floors for several days, the walnuts are mechanically swept and transferred for washing, mechanical de-hulling and artificial drying of in-shell walnuts with hot air (HA) heating [38]. Typically, walnuts are dried at around 43°C (110°F) until the MC of the walnuts are below 8% (w.b.) in the bin dryers (**Figure 3**).

Figure 3.
Typical harvesting and postharvest processing of walnuts: (A) walnut shaking; (B) hulling and washing section; (C) hot air drying bins.

This process can take as long as 24 h [39]. This conventional heated air drying method has the advantages of large processing capacity and relatively low operating cost [2]. Currently, there are three major concerns with this process: (1) the efficiency and throughput of the current drying method may not fulfill the growing production volume, resulting in significant product loss due to insufficient or inappropriate drying [5]; (2) walnut drying is very energy-intensive [2, 40] and causes large amount of carbon emission, thus efficient drying methods are needed to improve the sustainability; (3) freshly harvested walnuts have wide distribution of IMCs [41]. Drying the walnuts with different MCs together leads to over-dried or under-dried products, causing quality deterioration, food safety risk, and energy waste [42]. Therefore, it is desirable to sort the walnuts with different MCs first based on MCs and conduct the drying separately to improve moisture uniformity in products and avoid problems from current drying methods.

Pistachios are commonly harvested off-ground and dried artificially (**Figure 4**). Pistachio nuts have hard and naturally enclosed shell, which significantly restrict the moisture transfer rate, and natural drying is not popular [44]. In a typical process, pistachios are firstly de-hulled, washed, and dried with HA to pop-open the shells. A second hulling process is then used to remove the remaining hulls to obtain the in-shell pistachios. After that, the open-shell nuts and closed-shell nuts are separated with a rotating sieve [45]. After removing the remaining foreign

Figure 4.
Harvesting and post-harvest processing of pistachios in California: (A) off-ground harvesting [43]; (B) mechanical dehulling; (C) cylindrical hot air drying.

materials, defected or stained nuts, the pistachios may be further dried in batch or continuous dryers with HA the kernel reaches 5% MC to ensure the safety of the dried nuts [46–48].

3. Drying technologies

Current drying methods have drawn increasing concerns due to low efficiency, high energy consumption, large carbon footprint and risks of causing quality and safety deterioration. As the results, the tree nut industries are under the pressure to develop more efficient and sustainable drying technologies that can preserve the quality and safety of dried products. In recent years, research works have been conducted mainly in two aspects: (1) to improve or optimize the operating conditions of the current HA drying practice; (2) to develop novel drying technologies based on thermal and non-thermal methods.

3.1 Improve the hot air drying process

Elevating the drying temperature is a potential approach to reduce the drying time and improve the efficiency of the current HA drying practices. However, keeping the nuts at high temperatures for long times is not desirable, as it may induce significant quality deterioration to nuts, such as lipid oxidation and browning [23, 49]. Enhancing the drying rates by intensifying the heat and moisture transfer during the drying while maintaining the product qualities has become the key approach. Since the hull and shell have much higher IMCs than the kernel [2, 42], most thermal energy in the heated air is consumed for raising the temperature and evaporating moisture in the hull and shell, particularly in initial drying stages [50, 51]. As the results, the temperature increase in kernels is significantly slowed down, and the moisture removal from kernel to the environment is greatly restricted due to the reverse moisture gradient from the hull to the kernel [50]. Based on these characteristics, Chen [15] developed a new drying strategy by using high temperature heating in the beginning of drying to quickly heat up and partially dry the hull and shell; at the same time, due to the relatively low thermal conductivity of the kernel and shell, the kernels should not be over-heated; then before kernels reached a temperature that was high enough to cause significant oil quality deterioration, drying temperature was decreased to finish the drying. With the aids of experimental studies and mathematical modeling of the heat and moisture transfer (**Figure 5**) during the drying, the feasibility of this new drying strategy was verified [51, 52]. Drying in-shell walnuts by HA with step-down temperatures reduced the drying time and energy consumption by up to 40% and 24%, respectively, and obtained similar oil quality and kernel color in the dried products compared to the conventional practice. Similarly, using HA drying with step-down temperature and tempering for in-hull almonds significantly reduced the drying time and did not affect the quality of the dried almonds in terms of oil quality or incidence of concealed damage [13].

Sorting the nuts into different groups based on IMCs first and then drying them separately is another strategy of improving the drying efficiency, moisture uniformity and product quality. Khir et al. [42] studied the correlation between the MCs of walnuts and terminal velocities and developed a sorting method called the 'air knife'. The walnuts were separated into low and high MC groups and dried separately, which resulted in 18–28% energy saving compared to without sorting. Meanwhile, the uniformity of moisture in the dried products was greatly improved. This technology has been commercialized and installed in-line in walnut hulling and drying facilities in California (**Figure 6**). Similarly, the correlations between

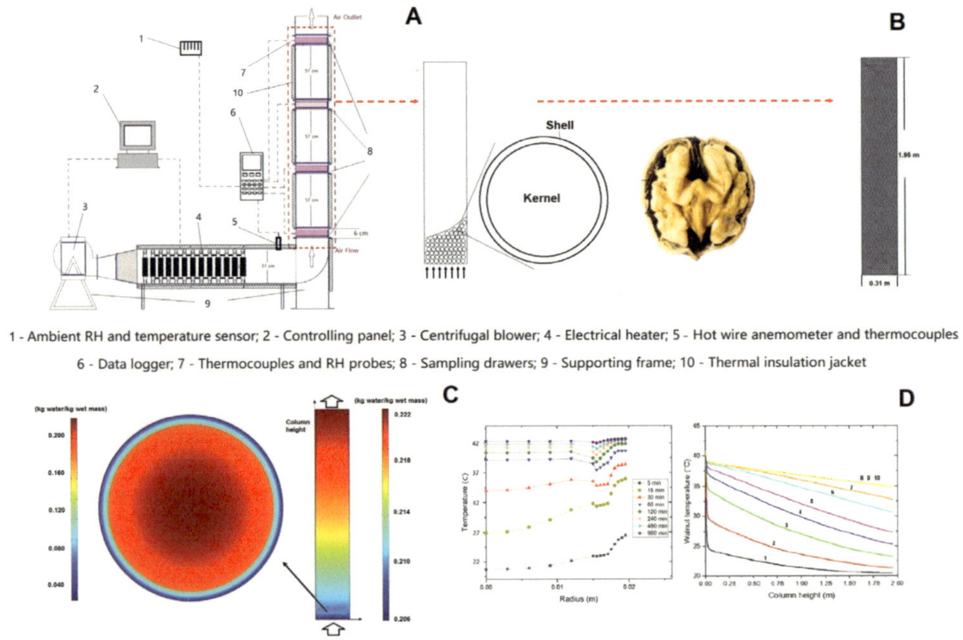

Figure 5.
Modeling results showing: (A) schematic diagram of pilot-scale column dryer; (B) mesh grid of the modeled system; (C) distribution of moisture and (D) temperature profile within the column and within single walnuts the moisture and temperature distribution within single walnut located in a pilot-scale column dryer [51, 52].

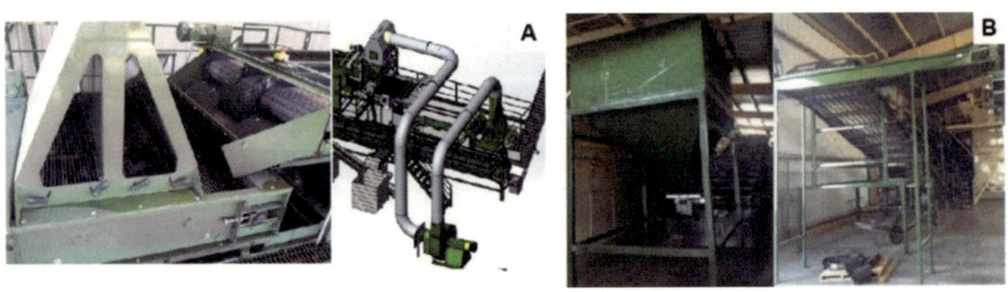

Figure 6.
Photo of the (A) 'air knife' sorter and (B) SIRHA dryer for walnuts.

terminal velocities of pistachios and almonds with their MCs have also been studied [44, 53]. Chen et al. [13, 54] have shown that in-hull almonds, in-shell almonds and loose hulls at harvest could be separated based on the thickness and terminal velocities of different groups, and suggested sorting and removing almond hulls prior to drying benefited the improvement of moisture uniformity and energy saving.

3.2 Novel thermal and non-thermal drying technologies

Infrared (IR) heating has the advantage of high heating intensity compared to convective heating. IR radiation could penetrate 2–5 mm depth into food surface [55], which matches the typical thickness of the nut hull and shell. Therefore, IR heating was an ideal technology to pre-dry the tree nuts. A commercial-scale sequential infrared and hot air (SIRHA) drying technology (**Figure 6**) was developed by Chen [15] and Atungulu et al. [40], which had 14.2 ton/h throughput and achieved 13.6–26.5% drying time reduction and 10–20% energy savings, respectively. Venkitasamy et al. [22, 23] used SIRHA drying for pistachios and almonds, and achieved 9% and 40%, respectively, compared with HA drying only. Additionally, research has shown that the percentage of shell splitting for pistachios increased with drying rates, and thus spray-rinsing with water and then IR drying could effectively improve the shell splitting [56].

Heat pump drying utilizes the thermal energy in the air flowing out of the dryer by dehumidifying and condensing the water vapor containing in the outflow air, retrieving the enthalpy of heat evaporation, then circulating the heat back to the dryer inlet [57]. HA heat pump drying was used for walnuts and higher drying efficiency was achieved without affecting the product quality [58, 59]. IR-assisted heat pump drying for walnuts reduced 20% drying time and 10% energy consumption compared to HA drying [60]. Solar heat pump drying reduced the drying time and increased the energy utilization ratio of pistachios [61, 62].

Dielectric heating such as microwave (MW, wavelength range: 300 MHz–300 GHz) and radiofrequency (RF, wavelength range: 3 kHz–300 MHz) can penetrate the foods. The electromagnetic wave activates the water molecules through dipole rotation and/or ionic polarization, generating heat volumetrically within foods, while non-polar molecules are not affected [63]. HA drying at 50°C assisted with RF heating at 27.12 MHz and 6 kW reduced 58.3% drying time of walnuts compared to HA drying only [64]. Intermittent MW drying could be used to dry pistachios without affecting the nut quality [65].

Some non-thermal technologies have also been studied as assistance to conventional drying processes. For example, high-power ultrasound treatments improved the drying rate and energy efficiency of pistachios [66]. The drying time of almonds was reduced by 58.33% with 40 min ultrasound treatment [67]. Such phenomena should be attributed to the enhancement of moisture transfer by the molecule vibrations induced by ultrasonic field. Under vacuum, the boiling point of water and vapor pressure in the drying chamber were significantly lower, and thus increase the driving force of moisture transfer during the drying process [12]. Vacuum drying was able to dry walnuts in a shorter time compared with conventional practice [68]. IR- and MW-assisted vacuum drying improved the drying efficiency of almonds [69].

4. Food safety processing: disinfection and disinfestation

Tree nuts are vulnerable to contamination by pathogenic microbes, such as *Salmonella* species, *Escherichia coli* strains and *Aspergillus flavus*, as well as molds and mildew [70, 71]. Two severe worldwide outbreaks of *Salmonella* occurred in 2001 and 2004 that were traced back to California almonds, which caused more than 200 illness cases in more than 15 US states or countries [72]. In response, ABC and United States Department of Agriculture (USDA) required all raw almonds to be pasteurized with at least 4-log reduction in the *Salmonella* population [73]. Additionally, aflatoxins are usually generated associated with the mold growth, which cause even more severe food safety risks. Insect damages (webbing, cast skins, frass, etc.) by field pests, such as Codling moth, (*Cydia pomonella* [*L.*]), navel orange worm (*Amyelois transitella [Walker]*), and storage pests, such as Red flour beetle and Indianmeal moth (*Plodia interpunctella [Hübner]*) have raised additional food safety concerns [74]. For example, navel orange worms lay eggs while the almonds are on tree, and hatch while they are dried on-ground. The insects could hide in dried nuts for a long time. In recent years, damaged nuts and live insects have been found in dried nuts or nut-containing products [35]. Insect infestation is also favorable for mold growth [75]. Although drying reduces water activity of nuts and can inhibit the microbial growth, drying alone is not enough to sufficiently disinfect and disinfest the tree nuts [13]. Thus, additional disinfection processing is needed to ensure the microbial safety of the products.

4.1 Conventional disinfection and disinfestation technologies

Conventionally disinfection and disinfestation technologies can be classified into chemical and thermal treatments, which are similar for different tree nuts. Chemical treatments refer to fumigation. The commonly used disinfectants and pesticides for fumigation of tree nuts include sulfuryl fluoride, aluminum phosphide and magnesium phosphide [74, 76]. Nowadays, there are increasing concerns of the chemical residues on tree nuts after fumigation as they are classified as 'probably carcinogenic to human', and excessive chemical use is not desired for the clean labeling of the products.

Thermal treatment is another type of disinfection and disinfestation method for tree nuts. High temperature heating kills the pathogenic microorganisms by denaturing their proteins and nucleic acids. For example, hot water blanching at 88°C for 1.6 min reduced 4-log *Salmonella* levels on the surface of almond kernels [73]. Oil roasting at 127°C for 53 s reduced 4-log *Salmonella* level on almonds and achieved effective disinfection of pistachios and walnuts [77].

4.2 Emerging disinfection and disinfestation technologies

In recent years, many emerging technologies have been studied and developed to improve the safety of tree nuts. It is generally agreed that 'moist heating' at high temperature and humidity conditions are effective for pasteurizing foods due to the high heat capacity and penetration depth. In addition, since 'moist heating' does not use harmful chemicals, it is more environmentally friendly and considered safer [78, 79]. Chen et al. [13, 37] applied step-down temperature HA heating with tempering for the simultaneous drying and disinfection of in-hull almonds that were harvested off-ground. When the average processing temperature was higher than 50°C, the insect eggs in the almonds were completely killed as no new larva was observed during the incubation and storage of dried nuts.

Radiative heating technologies have been proven to effectively pasteurize and disinfest the nuts. For example, IR preheating followed by temperature holding achieved 7.5-log reduction of *Salmonella enterica* on almonds owing to the intensive thermal effects [80]. SIRHA heating pasteurized pistachios and almonds with up to 6.1 and 4.7 log CFU/nut reductions in *Salmonella* level, respectively [22, 23]. MW and RF heating can penetrate the nuts and achieve rapid volumetric heating. As the results, 2–4 min of RF treatment was able to reduce 5-log *Salmonella* in the almonds [81]. RF heating could also kill the larvae of *Ceraphron cephalonica* and rice moth in walnuts [82], and *Indianmeal* moth in pistachios [83]. Nonetheless, due to the high temperature, thermal processing may damage the quality and reduce the shelf life of tree nuts, particularly causing lipid oxidation. Therefore, non-thermal disinfection technologies are gaining increasing interests.

Irradiation technologies have been used for disinfection for tree nuts. Ultraviolet (UV) illumination destructs the DNA structure of microorganisms and degrades the toxins. Particularly, Aflatoxin B1 (AFB1) absorbs UV radiation strongly at wavelengths of 362 nm and degrades rapidly under acidic ($pH < 3$) or alkaline ($pH > 10$) conditions [84]. Pulsed light treatment illuminates high intensity UV and/or visible light to target foods, which can also destruct the DNA structure in microbials and leads to effective disinfection [85]. Electron beam irradiation and Gamma irradiations have also been applied to disinfect tree nuts effectively [73, 86]. Low pressure cold plasma can generate UV radiation and reactive chemical species that destroy the protein and DNA structure within bacteria and fungi that infect the nuts [87]. However, it has also been reported that increased irradiation dose caused decrease in the nutrient contents in tree nuts, such as α-tocopherol [88, 89]. Therefore, these irradiation disinfection technologies need further investigation to maintain both the safety and quality of tree nuts.

Some other emerging technologies, such as nanoparticles (NPs), electrolyzed oxidized water (EOW) and ozone treatments have also been researched. Photocatalytic NPs can be used as a disinfectant alone or combined with other materials due to the oxidative stress arising from reactive oxygen species (ROS) that are generated under visible or UV lights [90]. Among which, silver NPs show antimicrobial properties against several bacteria including *Escherichia*, *Pseudomonas*, *Salmonella*, *Bacillus*, *Clostridium*, *Enterococcus*, *Listeria*, and *Streptococcus* [91], and different fungus, including *Aspergillus niger*, *Candida albicans*, and *Saccharomyces cerevisiae* [92, 93]. Some metal oxide photocatalytic NPs, such as titanium dioxide (TiO_2), also show potential disinfection functions [94, 95]. However, excessive use of NPs has raised concerns for their accumulation in foods that may cause toxicity.

EOW is normally obtained by passing the saltwater solution through an electrolysis system containing a cathode, an anode and a selective-permeable membrane [96]. The electrolysis of saltwater generates oxidizing species, such as O_2 and Cl_2 gases, and HOCl, at the anode. The redox potential of the EOW solution ranged from +700 to +800 mV with a pH of 4, which indicates high oxidizing ability [97]. These oxidizing species damage the cell wall and the metabolic process of a variety of pathogenic bacteria, such as *E. coli O157:H7*, *Listeria monocytogenes, Bacillus cereus*, and *Salmonella typhimurium* [98, 99]. Although EOWs are accepted as an antiseptic agent in food production, their toxicity needs further investigation.

Ozone is another ROS that has been used for food safety improvement. As a strong oxidant, ozone destructs cell wall, cell membrane and other cell constitutions in microorganisms [100, 101]. Ozone could also effectively degrade mycotoxins and aflatoxins in foods by reacting with the alkene double bonds [102]. Ozone is relatively unstable, and can spontaneously decompose into oxygen, thus do not generate hazardous residues in foods [101]. However, ozone is a greenhouse gas with strong global warming potential, and thus its application and emission may need to be regulated. Although not yet reported, these emerging technologies also show potentials to be used for the disinfection of tree nuts, and thus more research works are needed.

5. Further processing

Depending on the types of final products, 'raw nuts' (dried and disinfested) may be sold directly or further processed with the methods of blanching, kernel peeling and roasting.

5.1 Blanching and kernel peeling

Tree nut kernels with bright and white color are appealing to the consumers and are typically required for producing meals or milk, or consuming raw in salads [103]. For this purpose, blanching and peeling of kernels are usually needed. The peel on the kernel surface, also known as 'pellicle', has dark color and usually has high contents of natural antioxidants, such as tannins, phenolics and flavonoids

[104]. They protect the kernel from natural oxidations and show good antibacterial activities [105]. However, due to the abundant antioxidants and fiber, pellicles usually have bitter taste, high chewiness, and low solubility in drinks, which lowers the sensory satisfaction and affect the palatability [106]. Kernel peeling is mainly accomplished by physical or chemical methods.

Physical peeling involves blanching and mechanical abrasive peeling [107] and is the recommended method in the almond industry [108]. However, abrasive peeling is not recommended for walnut kernels since walnut kernel has irregular shape and abrasion may also result in significant loss of nut flesh. During blanching, nut kernels are subjected to either water soaking or steaming at high temperatures, which cause the pellicle to swell and crack [109]. After blanched, the tree nuts go through soft rubber rollers to mechanically peel off the pellicle by abrasion [106]. Besides, blanching also showed the capability to protect the color of the tree nut kernels [110], and control the cross-contamination of aflatoxins in almonds [111]. The disadvantages of blanching include the increase of MCs, softening of the nut flesh and leakage of polyphenols into the water [112], which brings up the need to further dry the kernels after peeling and loss of antioxidant activities. The large water consumption is another concern for the sustainability, particularly in the drought regions.

Chemical peeling usually refers to the hot lye peeling by NaOH, Na_2CO_3, and $Ca(OH)_2$, etc. [113, 114]. In a typical process, nut kernels are soaked in hot lye solutions for several minutes to corrode the pellicles, then rinsed with water [109]. Factors, such as alkali type, concentration, temperature, and soaking time, are important for the peeling performance [115]. Although lye peeling is efficient and effective, it may cause significant quality deteriorations, such as texture softening, surface browning, loss of crude protein and fat, decrease of antioxidant activities, and increased oxidation [113]. If the rinsing of peeled nuts was not performed adequately, the chemical residues on the nut surface cause additional food safety concerns. More importantly, disposal of the wastewater from lye peeling requires excessive use of chemicals and may cause severe environmental impacts.

For these reasons, novel and safe peeling methods have been studied and developed. IR radiation can penetrate the pellicle and heat up the kernel, which cause moisture evaporation and accumulation of vapor pressure under the pellicle. Meanwhile, IR heating may cause pyrolysis of pectin substances. When the vapor pressure reaches a critical level, the pellicles crack and can be peeled [109, 116]. Zhao et al. [117] developed a cryogenic peeling system for walnuts, in which the walnuts were held at −160°C by cold gas/liquid N_2 and moved dynamically downwards, and the shrinking pellicles were removed by upflowed air. Studies have shown that walnuts and almonds with their pellicles peeled off had shorter shelf life compared to the unpeeled ones [106, 118]. Therefore, applications of edible coatings containing antioxidant substances, such as Mastic gum, chitosan incorporated with green tea extract, walnut phenolic extracts, and mango kernel starch, etc. are popular research topics in recent years [118–121]. Meanwhile, the sensory quality of the nuts with edible coatings should not be compromised.

5.2 Roasting

Roasting is a commonly used processing to improve the palatability of tree nuts, which is usually done at a temperature higher than 90°C [122]. Maillard reaction between the carbonyl group of reducing sugars and the amino groups of proteins in the nuts is responsible for the nonenzymatic browning and formation of substances with desired 'roasted aroma and flavor' (e.g., pyrazines, furans, and pyrroles) [123–125]. During roasting, nuts are further dried and are subjected to some change in texture properties, which gives rise to the crunchy mouthfeel. The texture properties of tree nuts (hardness, fracture force, firmness, etc.) are significantly affected by the roasting temperatures [126, 127]. Meanwhile, roasting is also an effective measure to reduce the aflatoxin contents in tree nuts [128].

Conventionally, tree nuts are roasted by HA or oil [129]. HA roasting is usually performed at a temperature ranging from 100 to 180°C for up to 60 min [130]. The main problems with this method are the long processing time, high energy consumption and non-uniform roasting, which should be attributed to the slow convective and conductive heat transfer with HA heating. During oil roasting, tree nuts are immersed and fried in a vegetable oil at a temperature near 180°C for 7–8 min, followed by drying to remove the oil from surface [131]. Although the roasting process do not significantly reduce the contents of macronutrients, it also has some disadvantages. The major concern associated with roasting are the severe oxidation of polyunsaturated fatty acids that give rise to 'off-flavor' and the risk of toxin generation over the smoke point under the high temperature processing [132]. Some main chemical reactions related to the quality change during tree nut roasting are shown in **Figure 7**. In addition, oil roasting results in high oil intake in the products. Therefore, new roasting technologies need to be more efficient and cost effective, while not compromising the product quality and safety.

Formation of acrylamide, a group 2a carcinogen by WHO, from the free asparagine and reducing sugars in the nuts arises another food safety concern. Asadi et al. [133] found IR roasting caused the highest acrylamide content in pistachios, and MW roasting led to the lowest. Increasing the roasting temperature and time, and MW power facilitated acrylamide formation, since the formation rate of acrylamide increased with temperature [134]. Milczarek et al. [135] suggested that MW roasting was a promising method to replace the conventional HA roasting for almonds. Adding of salt during roasting can mitigate the acrylamide formation in tree nuts, which should be due to the prevention of intermediate (such as Schiff base) formation under the inhibition of cations [136]. In response, ABC [137] suggested that almonds should be roasted at the lowest possible temperature to minimize the acrylamide formation. Corradini and Peleg [138] found that using a step-down temperature heating may reduce the acrylamide contents in baked goods. Bagheri [139] suggested that use of IR heating together with HA or MW heating reduced the roasting time while obtaining similar product quality. Future research

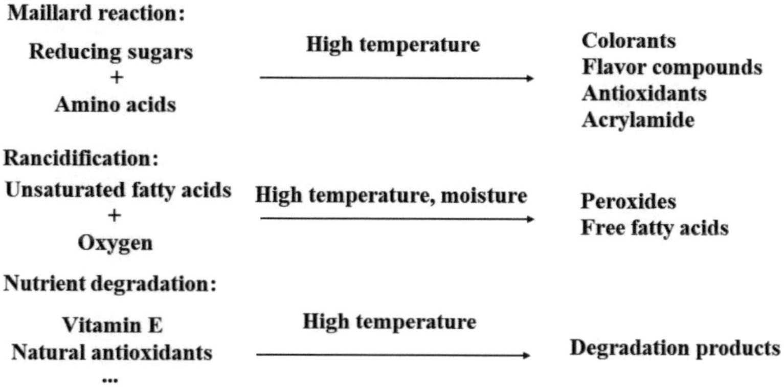

Figure 7.
Main chemical reactions related to quality change of nuts during roasting.

in tree nut roasting should focus on the development of novel roasting methods that combine the advantage of different heating methods, optimization of roasting parameters (using stepwise temperature roasting strategy) and addition of appropriate food-safe additives to preserve product quality, reduce acrylamide formation, and improve microbial safety and energy efficiency.

6. Utilization of tree nut processing wastes for value-added byproducts

Besides kernels, a large amount of waste materials, mainly the shells and hulls, is generated during tree nut processing and production. In fact, the tree nut hulls and shells account for 70% of the harvested weight [140], but currently have low economic values. Utilization of these wastes for producing value-added byproducts or bioenergy should benefit the stakeholders economically and reduce the environmental impacts of tree nut processing industry.

Hulls usually have high contents of natural antioxidants, such as tannins and phenolics [141]. Extraction of these antioxidants and then adding them as functional additives to other foods may improve the economic value of the crop [142]. Extracts from walnut and pistachio hulls exhibit good antibacterial and anti-inflammatory properties [143]. Walnut hull also contains juglone, a compound with high pharmaceutical value, which can be extracted and used in medicines [144]. Almond hulls weights over 60% at harvest and are conventionally used as cattle feed in California and Australia [145]. However, they are normally sold at a very low price. In addition, hulling and shelling of almonds cost about $0.30/kg, being one of the most expensive operations in the almond industry. Therefore, these processes need to be optimized to reduce the processing cost. Besides, off-ground harvested almonds have much higher MCs in the hulls at the time of harvest [54]. Wet dehulling of the almonds will be needed, as drying the hulls consumes more

Source	Valuable substance	Application	Reference
Walnut husk	Juglone and antioxidant	Enhance antioxidant and antimicrobial properties of ketchup	[141, 142]
Pistachio hull	Phenolic compound	Improve antibacterial and anti-inflammation properties	[143, 144]
Almond hull	Soluble sugar	Culture and produce edible fungi for foods	[148, 150]
Almond hull	Soluble sugar	Culture microorganisms that produce biodegradable plastics	[149]
Pistachio shell	Woody biomass	Biogas production	[146]
Almond hull	Soluble sugar and polysaccharides	Bioenergy production	[147]
Walnut and pistachio shell	Woody biomass	Biogas production	[152, 153]
Walnut shell	Cellulose	Reinforcement for biodegradable packaging	[154]
Pistachio processing waste	Starch, fiber and bioactive compounds	Biodegradable packaging for foods, active packaging	[155]

Table 1.
Valuable substances and potential applications of wastes and byproducts from tree nut processing.

than half of the total drying energy and leads to extra drying time and costs [24]. Meanwhile, since the high moisture hulls have high sugar contents, they can be used for fermentation and bioenergy production, and culturing certain microorganisms for producing value-added foods or biodegradable plastics without the need of drying [146–150]. Tree nut shells are mainly composed of cellulose, hemicellulose, and lignin [151], and thus can be used as woody biomass for biogas production through combustion or pyrolysis [152, 153]. The natural biopolymers from the tree nut production wastes and byproducts, such as polysaccharides, fibers, and proteins could also be utilized and manufactured into thin films as biodegradable packaging for foods [154]. The natural bioactive compounds in the hulls can be added into the packaging to improve the antioxidative and antimicrobial functions [155]. **Table 1** provides a summary of the potential values and applications of tree nut production wastes and byproducts.

7. Conclusion

Tree nuts are healthy foods with significant economic values. Postharvest processing technologies are essential in preserving the quality and safety of tree nuts. In this chapter, the status, recent advances, and challenges of drying, disinfection, disinfestation, blanching, peeling, and roasting of almonds, walnuts, and pistachios in scientific research, as well as in industrial productions are summarized. Current processing practices can be improved by optimizing the operating parameters. Novel thermal and non-thermal drying technologies, such as IR, MW, RF, ultrasound, vacuum, etc. and combined technologies, have shown great potential to improve the drying efficiency and safety of nuts without compromising the nut quality. It was noted that the postharvest processing methods of different tree nuts share some similarities but are also different due to the differences in physical properties and chemical compositions. Each processing practice has significant impacts on the quality and safety of the final products, such as lipid oxidation, loss of nutrients and formation of acrylamide. Therefore, suitable processing technologies and operating conditions must be carefully selected. Additionally, processing wastes of tree nuts, such as hulls and shells, have potential to be utilized as high value bioresources for producing value-added products. Being important economic crops with large market size, improvements in the postharvest processing technologies of tree nuts could translate into significant global influence in reducing the energy consumption and environmental impacts for improved sustainability, economic values, and competitiveness of tree nut industry.

Author details

Chang Chen and Zhongli Pan*
University of California Davis, Davis, United States of America

*Address all correspondence to: zlpan@ucdavis.edu

References

[1] Ayadi M, Ghrab M, Gargouri K, Elloumi O, Zribi F, Ben Mimoun M, et al. Kernel characteristics of almond cultivars under rainfed conditions. Acta Horticulturae. 2006;**726**:377-381

[2] Chen C, Weipeng Z, Venkitasamy C, Khir R, McHugh T, Pan Z. Walnut structure and its influence on the hydration and drying characteristics. Drying Technology. 2020;**38**(8):975-986

[3] Intrado GlobeNewswire. Global Edible Nuts Industry [Internet]. 2020. Available from: https://www.globenewswire.com/news-release/2020/07/23/2066759 /0/en/Global-Edible-Nuts-Industry.html [Accessed: 01 December 2021]

[4] WorldAtlas. The most popular nuts in the world [Internet]. 2021. Available from: https://www.worldatlas.com/articles/the-most-popular-nuts-in-the-world.html [Accessed: 01 December 2021]

[5] USDA, National Agricultural Statistics Service. 2020 California Walnut Objective Measurement Report [Internet]. 2020. Available from: https://www.nass.usda.gov/Statistics_by_State/California/Publications/Specialty_and_Other_Releases/Walnut/Objective-Measurement/202008walom.Pdf [Accessed: 01 December 2021]

[6] Chen C, Pan Z. Postharvest processing of tree nuts: Current status and future prospects – A comprehensive review. Critical reviews in food science and food safety. 2021. Accepted for publication

[7] Labuza TP, McNally L, Gallagher D, Hawkes J, Hurtado F. Stability of intermediate moisture foods. 1. Lipid oxidation. Journal of Food Science. 1972;**37**(1):154-159

[8] Khir R, Pan Z, Atungulu GG, Thompson JF. Characterization of physical and aerodynamic properties of walnuts. Transactions of the ASABE. 2014;**57**(1):53-61

[9] Virtanen JK, Mursu J, Voutilainen S, Uusitupa M, Tuomainen TP. Serum omega-3 polyunsaturated fatty acids and risk of incident type 2 diabetes in men: The Kuopio ischemic heart disease risk factor study. Diabetes Care. 2014;**37**(1):189-196

[10] Gecgel U, Gumus T, Tasan M, Daglioglu O, Arici M. Determination of fatty acid composition of γ-irradiated hazelnuts, walnuts, almonds, and pistachios. Radiation Physics and Chemistry. 2011;**80**(4):578-581

[11] Chen C, Venkitasamy C, Zhang W, Deng L, Meng X, Pan Z. Effect of step-down temperature drying on energy consumption and product quality of walnuts. Journal of Food Engineering. 2020;**285**:110105

[12] Zhang WP, Chen C, Pan Z, Xiao HW, Xie L, Gao ZJ, et al. Design and performance evaluation of a pilot-scale pulsed vacuum infrared drying (PVID) system for drying of berries. Drying Technology. 2020;**38**(10):1340-1355

[13] Chen C, Khir R, Shen Y, Wu X, Zhang R, Cao X, et al. Energy consumption and product quality of off-ground harvested almonds under hot air column drying. LWT Food Science and Technology. 2021;**138**:110768

[14] Zhang W, Pan Z, Xiao H, Zheng Z, Chen C, Gao Z. Pulsed vacuum drying (PVD) technology improves drying efficiency and quality of Poria cubes. Drying Technology. 2018;**36**(8):908-921

[15] Chen C. Characteristics and mechanisms of walnut drying under hot air and infrared heating [doctoral

dissertation]. Davis: University of California

[16] Zhang W, Chen C, Pan Z, Zheng Z. Vacuum and infrared-assisted hot air impingement drying for improving the processing performance and quality of Poria cocos (Schw.) wolf cubes. Food. 2021;**10**(5):992

[17] Tavakolipour H. Postharvest operations of pistachio nuts. Journal of Food Science and Technology. 2015;**52**(2):1124-1130

[18] Ajith S, Pramod S, Kumari CP, Potty VP. Effect of storage temperatures and humidity on proximate composition, peroxide value and iodine value of raw cashew nuts. Journal of Food Science and Technology. 2015;**52**(7):4631-4636

[19] Almond Board of California. Almond Almanac [Internet]. 2015. Available from: https://www.almonds.com/sites/default/files/content/attachments/2015_almanac.pdf [Accessed: 01 December 2021]

[20] Harris LJ, Yada S, Beuchat LR, Danyluk MD. Outbreaks of foodborne illness associated with the consumption of tree nuts, peanuts, and sesame seeds (version 2). In Outbreaks from tree nuts, peanuts, and sesame seeds [Internet]. 2021. Available from: https://ucfoodsafety.ucdavis.edu/low-moisture-foods/nuts-and-nut-pastes [Accessed: 01 December 2021]

[21] Farakos SM, Pouillot R, Johnson R, Spungen J, Son I, Anderson N, et al. A quantitative assessment of the risk of human salmonellosis arising from the consumption of almonds in the United States: The impact of preventive treatment levels. Journal of Food Protection. 2017;**80**(5):863-878

[22] Venkitasamy C, Brandl MT, Wang B, McHugh TH, Zhang R, Pan Z. Drying and decontamination of raw pistachios with sequential infrared drying, tempering and hot air drying. International Journal of Food Microbiology. 2017;**246**:85-91

[23] Venkitasamy C, Zhu C, Brandl MT, Niederholzer FJ, Zhang R, McHugh TH, et al. Feasibility of using sequential infrared and hot air for almond drying and inactivation of *Enterococcus faecium* NRRL B-2354. LWT Food Science and Technology. 2018;**95**:123-128

[24] Chen C, Abd El Gebreil RK, Shen Y, Wu X, Ning Z, Niederholzer F, et al. Energy and quality analyses of off-ground harvested almonds under hot air column drying. In: 2020 ASABE Annual International Virtual Meeting. American Society of Agricultural and Biological Engineers. 2020. p. 1

[25] Ponpesh P, Giles DK, Downey D. Mitigation of in-orchard dust through modified harvester operation. Transactions of the ASABE. 2010;**53**(4):1037-1044

[26] Ritchie H. Food production is responsible for one-quarter of the world's greenhouse gas emissions [Internet]. 2019. Available from: https://ourworldindata.org/food-ghg-emissions [Accessed: 01 December 2021]

[27] Energy Information Administration, EIA. International Energy Outlook 2017 [Internet]. 2017. Available from: https://www.eia.gov/outlooks/archive/ieo17/pdf/exec_summ.pdf [Accessed: 01 December 2021]

[28] Bajan B, Mrówczyńska-Kamińska A, Poczta W. Economic energy efficiency of food production systems. Energies. 2020;**13**(21):5826

[29] Schreyer F, Luderer G, Rodrigues R, Pietzcker RC, Baumstark L, Sugiyama M, et al. Common but differentiated leadership: Strategies and challenges for carbon neutrality by 2050 across industrialized economies.

Environmental Research Letters. 2020;**15**(11):114016

[30] What do we need to achieve at COP26? [Internet]. 2021. Available from: https://ukcop26.org/cop26-goals/ [Accessed: 01 December 2021]

[31] Brown, B. Australian almond industry study tour of Spain. Almond Board of Australia (ABA) [Internet]. 2021. Available from: https://www.horticulture.com.au/globalassets/laserfiche/assets/project-reports/al13701/al13701-final-report.pdf-71.pdf [Accessed: 01 December 2021]

[32] Almond Board of Australia. 2019/2020 Annual Report [Internet]. 2020. Available from: https://australianalmonds.com.au/wp-content/uploads/2020/12/ABA_2020_AnnualReport_low_res.pdf [Accessed: 01 December 2021]

[33] Mayanja IK, Coates MC, Niederholzer F, Donis-Gonzales IR. Development of a stockpile heated and ambient air dryer (shad) for freshly harvested almonds. Applied Engineering in Agriculture. 2021;**37**(3):417-425

[34] Baticados EJ, Capareda SC, Maglinao AL. Particulate matter emission factors using low-dust harvesters for almond nut-picking operations. Journal of the Air & Waste Management Association. 2019;**69**(11):1304-1311

[35] Yu J, Ren S, Liu C, Wei B, Zhang L, Younas S, et al. Non-destructive detection and classification of in-shell insect-infested almonds based on multispectral imaging technology. The Journal of Agricultural Science. 2018;**156**(9):1103-1110

[36] Gradziel T, Mahoney N, Abdallah A. Aflatoxin production among almond genotypes is not related to either kernel oil composition or *Aspergillus flavus* growth rate. HortScience. 2000;**35**(5): 937-939

[37] Chen C, Liao C, Wongso I, Wang W, Khir R, Huang G, et al. Drying and disinfection of off-ground harvested almonds using step-down temperature hot air heating. LWT Food Science and Technology. 2021;**152**:112282

[38] Chen C, Liao C, Wongso I, Wang W, Abd El Gebreil RK, Ning Z, Huang G, Niederholzer F, Wang L, Pan Z. Drying and disinfection of off-ground harvested almonds using step-down temperature hot air heating. In: 2021 ASABE Annual International Virtual Meeting. American Society of Agricultural and Biological Engineers. 2021. p. 1

[39] Batchelor LD. Walnut Culture in California. Berkeley, USA: California Agricultural Experiment Station; 1924

[40] Atungulu GG, Teh HE, Wang T, Fu R, Wang X, Khir R, et al. Infrared pre-drying and dry-dehulling of walnuts for improved processing efficiency and product quality. Applied Engineering in Agriculture. 2013;**29**(6):961-971

[41] Khir R, Pan Z, Atungulu GG, Thompson JF, Shao D. Size and moisture distribution characteristics of walnuts and their components. Food and Bioprocess Technology. 2013;**6**(3): 771-782

[42] Khir R, Atungulu GG, Pan Z, Thompson JF, Zheng X. Moisture-dependent color characteristics of walnuts. International Journal of Food Properties. 2014;**17**(4):877-890

[43] Rosa UA, Rosenstock TS, Choi H, Pursell D, Gliever CJ, Brown PH, et al. Design and evaluation of a yield monitoring system for pistachios. Transactions of the ASABE. 2011;**54**(5):1555-1567

[44] Polat R, Gezer I, Guner M, Dursun E, Erdogan D, Bilim HC. Mechanical harvesting of pistachio nuts.

Journal of Food Engineering. 2007;**79**(4):1131-1135

[45] Taher Agroindustrial Group. Video: Fresh Pistachio Processing Line [Internet]. 2018. Available from: https://www.youtube.com/watch?v=yrIKzKbz73Y [Accessed: 01 December 2021]

[46] Kashani Nejad M, Tabil LG, Mortazavi A, Safe KA. Effect of drying methods on quality of pistachio nuts. Drying Technology. 2003;**21**(5):821-838

[47] Gazor HR, Minaei S. Influence of temperature and air velocity on drying time and quality parameters of pistachio (*Pistacia vera* L.). Drying Technology. 2005;**23**(12):2463-2475

[48] Shakerardekani A, Karim R, Ghazali HM, Chin NL. Types of dryers and their effect on the pistachio nuts quality—A review. 2011. SSRN: 1965354

[49] Mu H, Gao H, Chen H, Fang X, Zhou Y, Wu W, et al. Study on the volatile oxidation compounds and quantitative prediction of oxidation parameters in walnut (*Carya cathayensis* Sarg.) oil. European Journal of Lipid Science and Technology. 2019;**121**:6, 1800521

[50] Chen C, Venkitasamy C, Zhang W, Khir R, Upadhyaya S, Pan Z. Effective moisture diffusivity and drying simulation of walnuts under hot air. International Journal of Heat and Mass Transfer. 2020;**150**:119283

[51] Chen C, Upadhyaya S, Khir R, Pan Z. Simulation of walnut drying under hot air heating using a nonequilibrium multiphase transfer model. Drying Technology. 2020:1-5. DOI: 10.1080/07373937.2020.1846552

[52] Chen C, Pan Z. Heat and moisture transfer studies on walnuts during hot air drying in a fixed-bed column dryer. Applied Thermal Engineering. 2021;**199**:117554

[53] Aydin C. Physical properties of almond nut and kernel. Journal of Food Engineering. 2003;**60**(3):315-320

[54] Chen C, Khir R, Zhang R, Cao X, Ning Z, Shen Y, et al. Development of sorting methods based on physical and aerodynamic properties of off-ground harvested almonds. International Journal of Agricultural and Biological Engineering. 2021;**14**(2):218-225

[55] Pan Z, Atungulu GG. Infrared Heating for Food and Agricultural Processing. Boca Raton, USA: CRC Press; 2010

[56] Nazari M, Ghanbarian D, Shakerardekani A, Maleki A. Investigating different methods of closed Shell pistachios splitting and effects of freezing prior to drying on Shell splitting percentage. Journal of Nuts. 2017;**8**(02):107-114

[57] Walmsley TG, Klemeš JJ, Walmsley MR, Atkins MJ, Varbanov PS. Innovative hybrid heat pump for dryer process integration. 2017;**57**:1039-1044

[58] Chen YX. Study on Walnut Drying by Air Source Heat Pump Combined with Solar Energy System. Master Degree Thesis. Kunming, China: Yunnan Normal University; 2015

[59] Chen YX, Lan Q, Ji X, Peng FM, Xia CF, Li XY. Experimental study on heat pump drying characteristics for walnuts. Journal of Yunnan Normal University (Natural Science Edition). 2014;**4**:30-34

[60] Dolgun GK, Aktaş M, Dolgun EC. Infrared convective drying of walnut with energy-exergy perspective. Journal of Food Engineering. 2021;**306**:110638

[61] Mokhtarian M, Tavakolipour H, Ashtari AK. Effects of solar drying along with air recycling system on physicochemical and sensory properties of dehydrated pistachio nuts. LWT Food

Science and Technology. 2017;**75**: 202-209

[62] Mokhtarian M, Tavakolipour H, Kalbasi-Ashtari A. Energy and exergy analysis in solar drying of pistachio with air recycling system. Drying Technology. 2016;**34**(12):1484-1500

[63] Ramaswamy H, Tang J. Microwave and radio frequency heating. Food Science and Technology International. 2008;**14**(5):423-427

[64] Zhang B, Zheng A, Zhou L, Huang Z, Wang S. Developing hot air-assisted radio frequency drying for in-shell walnuts. Emirates Journal of Food and Agriculture. 2016;**7**(1):459-467

[65] Kermani AM, Khashehchi M, Kouravand S, Sadeghi A. Effects of intermittent microwave drying on quality characteristics of pistachio nuts. Drying Technology. 2017;**35**(9):1108-1116

[66] Kouchakzadeh A. The effect of acoustic and solar energy on drying process of pistachios. Energy Conversion and Management. 2013;**67**:351-356

[67] Kaveh M, Jahanbakhshi A, Abbaspour-Gilandeh Y, Taghinezhad E, Moghimi MB. The effect of ultrasound pre-treatment on quality, drying, and thermodynamic attributes of almond kernel under convective dryer using ANNs and ANFIS network. Journal of Food Process Engineering. 2018;**41**(7): e12868

[68] Zhou X, Gao H, Mitcham EJ, Wang S. Comparative analyses of three dehydration methods on drying characteristics and oil quality of in-shell walnuts. Drying Technology. 2018;**36**(4):477-490

[69] Safary M, Chayjan RA. Optimization of almond kernels drying under infrared vacuum condition with microwave pretreatment using response surface method and genetic algorithm. 2016;**18**:1543-1556

[70] Uesugi AR, Harris LJ. Growth of Salmonella enteritidis phage type 30 in almond hull and shell slurries and survival in drying almond hulls. Journal of Food Protection. 2006;**69**(4):712-718

[71] Blessington T, Mitcham EJ, Harris LJ. Survival of *Salmonella enterica*, *Escherichia coli* O157: H7, and *Listeria monocytogenes* on inoculated walnut kernels during storage. Journal of Food Protection. 2012;**75**(2):245-254

[72] Feng Y, Lieberman VM, Jung J, Harris LJ. Growth and survival of foodborne pathogens during soaking and drying of almond (*Prunus dulcis*) kernels. Journal of Food Protection. 2020;**83**(12):2122-2133

[73] Cuervo MP, Lucia LM, Castillo A. Efficacy of traditional almond decontamination treatments and electron beam irradiation against heat-resistant *Salmonella* strains. Journal of Food Protection. 2016;**79**(3):369-375

[74] Mitcham EJ, Veltman RH, Feng X, de Castro ED, Johnson JA, Simpson TL, et al. Application of radio frequency treatments to control insects in in-shell walnuts. Postharvest Biology and Technology. 2004;**33**(1):93-100

[75] Johnson JA, Vail PV, Brandl DG, Tebbets JS, Valero KA. Integration of nonchemical treatments for control of postharvest pyralid moths (Lepidoptera: Pyralidae) in almonds and raisins. Journal of Economic Entomology. 2002;**95**(1):190-199

[76] Isikber AA, Navarro S, Finkelman S, Rindner M. Propylene oxide: A potential quarantine and pre-shipment fumigant for disinfestation of nuts. Phytoparasitica. 2006;**34**(4):412-419

[77] Pan Z, Zhang R, Zicari S, editors. Integrated Processing Technologies for Food and Agricultural by-products. Cambridge, USA: Academic Press; 2019

[78] Silva FV, Gibbs PA. Thermal pasteurization requirements for the inactivation of *Salmonella* in foods. Food Research International. 2012;**45**(2):695-699

[79] Villa-Rojas R, Tang J, Wang S, Gao M, Kang DH, Mah JH, et al. *Thermal inactivation of S. enteritidis* PT 30 in almond kernels as influenced by water activity. Journal of Food Protection. 2013;**76**(1):26-32

[80] Brandl MT, Pan Z, Huynh S, Zhu Y, McHugh TH. Reduction of *S. enteritidis* population sizes on almond kernels with infrared heat. Journal of Food Protection. 2008;**71**(5):897-902

[81] Gao M, Tang J, Villa-Rojas R, Wang Y, Wang S. Pasteurization process development for controlling *Salmonella* in in-shell almonds using radio frequency energy. Journal of Food Engineering. 2011;**104**(2):299-306

[82] Mao Y, Wang P, Wu Y, Hou L, Wang S. Effects of various radio frequencies on combined drying and disinfestation treatments for in-shell walnuts. LWT Food Science and Technology. 2021;**144**:111246

[83] Ling B, Hou L, Li R, Wang S. Storage stability of pistachios as influenced by radio frequency treatments for postharvest disinfestations. Innovative Food Science & Emerging Technologies. 2016;**33**:357-364

[84] Jubeen F, Bhatti IA, Khan MZ, Zahoor-Ul H, Shahid M. Effect of UVC irradiation on aflatoxins in ground nut (*Arachis hypogea*) and tree nuts (*Juglans regia*, *Prunus duclus* and *pistachio vera*). Journal of Chemical Society of Pakisthan. 2012;**34**(6):1366-1374

[85] Deng LZ, Tao Y, Mujumdar AS, Pan Z, Chen C, Yang XH, et al. Recent advances in non-thermal decontamination technologies for microorganisms and mycotoxins in low-moisture foods. Trends in Food Science & Technology. 2020;**106**:104-112

[86] Iuliano S, Livingstone K, Stonehouse W, Coates A. Abstracts of the 44th Annual Scientific Meeting of the Nutrition Society of Australia. Multidisciplinary Digital Publishing Institute Proceedings. 2021;**72**(1):1

[87] Ahangari M, Ramezan Y, Khani MR. Effect of low pressure cold plasma treatment on microbial decontamination and physicochemical properties of dried walnut kernels (*J. regia* L.). Journal of Food Process Engineering. 2021;**44**(1):e13593

[88] Di Stefano V, Pitonzo R, Bartolotta A, D'Oca MC, Fuochi P. Effects of γ-irradiation on the α-tocopherol and fatty acids content of raw unpeeled almond kernels (*P. dulcis*). LWT Food Science and Technology. 2014;**59**(1):572-576

[89] Iqbal M, Bhatti IA, Shahid M, Nisar J. Physicochemical characterization, microbial decontamination and shelf life analysis of walnut (*J. regia* L) oil extracted from gamma radiation treated seeds. Biocatalysis and Agricultural Biotechnology. 2016;**6**:116-122

[90] Li F, Liang Z, Zheng X, Zhao W, Wu M, Wang Z. Toxicity of nano-TiO_2 on algae and the site of reactive oxygen species production. Aquatic Toxicology. 2015;**158**:1-3

[91] Wijnhoven SW, Peijnenburg WJ, Herberts CA, Hagens WI, Oomen AG, Heugens EH, et al. Nano-silver—A review of available data and knowledge gaps in human and environmental risk assessment. Nanotoxicology. 2009;**3**(2): 109-138

[92] Marambio-Jones C, Hoek EM. A review of the antibacterial effects of silver nanomaterials and potential implications for human health and the environment. Journal of Nanoparticle Research. 2010;**12**(5):1531-1551

[93] Li B, Wang H, Zhang B, Hu P, Chen C, Guo L. Facile synthesis of one dimensional AgBr@ Ag nanostructures and their visible light photocatalytic properties. ACS Applied Materials & Interfaces. 2013;**5**(23):12283-12287

[94] Clemente Z, Castro VL, Moura MA, Jonsson CM, Fraceto LF. Toxicity assessment of TiO_2 nanoparticles in zebrafish embryos under different exposure conditions. Aquatic Toxicology. 2014;**147**:129-139

[95] Wei W, Gao S, Yang Z, Wu Y, Chen C, Guo L, et al. Porous SnO_2 nanocubes with controllable pore volume and their Li storage performance. RSC Advances. 2014;**4**(26):13250-13255

[96] Subrota H, Surajit M, PS M, Shilpa V, Yogesh K, Dipika Y. Electrolyzed oxidized water (EOW): Non-thermal approach for decontamination of food borne microorganisms in food industry. Food and Nutrition Sciences. 2012;**3**:760-768

[97] Kim C, Hung YC, Brackett RE, Frank JF. Inactivation of *L. monocytogenes* biofilms by electrolyzed oxidizing water. Journal of Food Processing and Preservation. 2001;**25**(2):91-100

[98] Kim C, Hung YC, Brackett RE, Lin CS. Efficacy of electrolyzed oxidizing water in inactivating *Salmonella* on alfalfa seeds and sprouts. Journal of Food Protection. 2003;**66**(2): 208-214

[99] Fabrizio KA, Cutter CN. Stability of electrolyzed oxidizing water and its efficacy against cell suspensions of *Salmonella typhimurium* and *L. monocytogenes*. Journal of Food Protection. 2003;**66**(8):1379-1384

[100] Deng LZ, Mujumdar AS, Pan Z, Vidyarthi SK, Xu J, Zielinska M, et al. Emerging chemical and physical disinfection technologies of fruits and vegetables: A comprehensive review. Critical Reviews in Food Science and Nutrition. 2020;**60**(15):2481-2508

[101] Deng LZ, Mujumdar AS, Zhang Q, Yang XH, Wang J, Zheng ZA, et al. Chemical and physical pretreatments of fruits and vegetables: Effects on drying characteristics and quality attributes—a comprehensive review. Critical Reviews in Food Science and Nutrition. 2019;**59**(9):1408-1432

[102] Subrota H, Surajit M, PS M, Shilpa V, Yogesh K, BP S, Dipika Y. Electrolyzed oxidized water (EOW): Non-thermal approach for decontamination of food borne microorganisms in food industry. Food and Nutrition Sciences. 2012;**64**(47): 8959-8972

[103] Wang W, Wang Q, Jin Y, Cheng A, Guo X, Liu C, et al. Study on the processing technology of instant walnut sauce. Food Research and Development. 2016;**37**(11):62-65

[104] Medic A, Jakopic J, Hudina M, Solar A, Veberic R. Identification and quantification of the major phenolic constituents in *J. regia* L. peeled kernels and pellicles, using HPLC–MS/MS. Food Chemistry. 2021;**352**:129404

[105] Zaini PA, Feinberg NG, Grilo FS, Saxe HJ, Salemi MR, Phinney BS, et al. Comparative proteomic analysis of walnut (*J. regia* L.) pellicle tissues reveals the regulation of nut quality attributes. Life. 2020;**10**(12):314

[106] Li C, Zhang S, Sun H, Zhao H, Chen C, Xing S. Study on a two-dimensional correlation visible–near infrared spectroscopy kinetic model for

the moisture content of fresh walnuts stored at room temperature. Journal of Food Process Engineering. 2020;**43**(12): e13551

[107] Shakerardekani A, Mohamadi A. Determination of peeling efficiency, free fatty acid, peroxide value and sensory evaluation of peeled pistachio kernel using hot water. Journal of Nuts. 2019;**10**(2):175-185

[108] Almond Board of California. Guidelines for Validation of Blanching Processes [Internet]. 2017. Available from: https://www.almonds.com/sites/default/files/content/attachments/blanching-validation-guidelines. pdf [Accessed: 01 December 2021]

[109] Liu M, Li C, Cao C, Wang L, Li X, Che J, et al. Walnut fruit processing equipment: Academic insights and perspectives. Food Engineering Reviews. 2021;**6**:1-36

[110] Bolling BW. Almond polyphenols: Methods of analysis, contribution to food quality, and health promotion. Comprehensive Reviews in Food Science and Food Safety. 2017;**16**(3):346-368

[111] Mahoney NE, Cheng LW, Palumbo JD. Effect of blanching on aflatoxin contamination and cross-contamination of almonds. Journal of Food Protection. 2020;**83**(12):2187-2192

[112] Hughey CA, Janusziewicz R, Minardi CS, Phung J, Huffman BA, Reyes L, et al. Distribution of almond polyphenols in blanch water and skins as a function of blanching time and temperature. Food Chemistry. 2012;**131**(4):1165-1173

[113] Wu S, Qin L, Jiang C, Zhang S. Dynamic changes of nutritional and functional components of walnut kernel during lye peeling. China Oils and Fats. 2013;**38**(2):84-87

[114] Li X, Pan Z, Atungulu GG, Zheng X, Wood D, Delwiche M, et al. Peeling of tomatoes using novel infrared radiation heating technology. Innovative Food Science & Emerging Technologies. 2014;**21**:123-130

[115] Niu ML, Xing JH, Zhang Q, Zhou JH. Study on the processing technology and formula of wolfberry and walnut milk. Hubei Agricultural Sciences. 2012;**12**:2549-2551

[116] Eskandari J, Kermani AM, Kouravand S, Zarafshan P. Design, fabrication, and evaluation a laboratory dry-peeling system for hazelnut using infrared radiation. LWT Food Science and Technology. 2018;**90**: 570-576

[117] Zhao Y, Chen L, Ji W, Guo J, Wang J. Study on a novel energy-saving cryogenic pre-treatment equipment for walnut kernel peeling. Food Control. 2021;**121**:107650

[118] Farooq M, Azadfar E, Rusu A, Trif M, Poushi MK, Wang Y. Improving the shelf life of peeled fresh almond kernels by edible coating with mastic gum. Coatings. 2021;**11**(6):618

[119] Sabaghi M, Maghsoudlou Y, Khomeiri M, Ziaiifar AM. Active edible coating from chitosan incorporating green tea extract as an antioxidant and antifungal on fresh walnut kernel. Postharvest Biology and Technology. 2015;**110**:224-228

[120] Grosso AL, Riveros C, Asensio CM, Grosso NR, Nepote V. Improving walnuts' preservation by using walnut phenolic extracts as natural antioxidants through a walnut protein-based edible coating. Journal of Food Science. Oct 2020;**85**(10):3043-3051

[121] Nawab A, Alam F, Haq MA, Lutfi Z, Hasnain A. Effect of mango kernel starch coatings on the shelf life of almond (*P. dulcis*) kernels. Journal of Food Processing and Preservation. 2018;**42**(2):e13449

[122] Sheikhshoaei H, Dowlati M, Aghbashlo M, Rosen MA. Exergy analysis of a pistachio roasting system. Drying Technology. 2020;**38**(12): 1565-1583

[123] Lin JT, Liu SC, Hu CC, Shyu YS, Hsu CY, Yang DJ. Effects of roasting temperature and duration on fatty acid composition, phenolic composition, Maillard reaction degree and antioxidant attribute of almond (*P. dulcis*) kernel. Food Chemistry. 2016;**190**:520-528

[124] Ling B, Yang X, Li R, Wang S. Physicochemical properties, volatile compounds, and oxidative stability of cold pressed kernel oils from raw and roasted pistachio (*P. vera* L. Var Kerman). European Journal of Lipid Science and Technology. 2016;**118**(9):1368-1379

[125] Rabadán A, Gallardo-Guerrero L, Gandul-Rojas B, Álvarez-Ortí M, Pardo JE. Effect of roasting conditions on pigment composition and some quality parameters of pistachio oil. Food Chemistry. 2018;**264**:49-57

[126] Nikzadeh V, Sedaghat N. Physical and sensory changes in pistachio nuts as affected by roasting temperature and storage. American-Eurasian Journal of Agricultural & Environmental Science. 2008;**4**(4):478-483

[127] Kahyaoglu T. Optimization of the pistachio nut roasting process using response surface methodology and gene expression programming. LWT Food Science and Technology. 2008;**41**(1): 26-33

[128] Yazdanpanah H, Mohammadi T, Abouhossain G, Cheraghali AM. Effect of roasting on degradation of aflatoxins in contaminated pistachio nuts. Food and Chemical Toxicology. 2005;**43**(7): 1135-1139

[129] Garcıa-Pascual P, Mateos M, Carbonell V, Salazar DM. Influence of storage conditions on the quality of shelled and roasted almonds. Biosystems Engineering. 2003;**84**(2):201-209

[130] Hojjati M, Lipan L, Carbonell-Barrachina ÁA. Effect of roasting on physicochemical properties of wild almonds (*Amygdalus scoparia*). Journal of the American Oil Chemists' Society. 2016;**93**(9):1211-1220

[131] Özdemir M, Devres O. Analysis of color development during roasting of hazelnuts using response surface methodology. Journal of Food Engineering. 2000;**45**(1):17-24

[132] Schlörmann W, Birringer M, Böhm V, Löber K, Jahreis G, Lorkowski S, et al. Influence of roasting conditions on health-related compounds in different nuts. Food Chemistry. 2015;**180**:77-85

[133] Asadi S, Aalami M, Shoeibi S, Kashaninejad M, Ghorbani M, Delavar M. Effects of different roasting methods on formation of acrylamide in pistachio. Food Science & Nutrition. 2020;**8**(6):2875-2881

[134] Süvari M, Sivri GT, Öksüz Ö. Effect of different roasting temperatures on acrylamide formation of some different nuts. IOSR Journal of Environmental Science, Toxicology and Food Technology. 2017;**11**(4):38-43

[135] Milczarek RR, Avena-Bustillos RJ, Peretto G, McHugh TH. Optimization of microwave roasting of almond (P runus dulcis). Journal of Food Processing and Preservation. 2014;**38**(3):912-923

[136] Kukurová K, Morales FJ, Bednarikova A, Ciesarova Z. Effect of L-asparaginase on acrylamide mitigation in a fried-dough pastry model. Molecular Nutrition & Food Research. 2009;**53**(12):1532-1539

[137] Almond Board of California. Acrylamide in Roasted Almonds

[Internet]. 2014. Available from: https://www.almonds.com/sites/default/files/content/attachments/aq0104_acrylamide_in_roasted_almonds.pdf [Accessed: 01 December 2021]

[138] Corradini MG, Peleg M. Linear and non-linear kinetics in the synthesis and degradation of acrylamide in foods and model systems. Critical Reviews in Food Science and Nutrition. 2006;**46**(6): 489-517

[139] Bagheri H. Application of infrared heating for roasting nuts. Journal of Food Quality. 2020;**4**:2020

[140] Almond Board of California. ALMOND TREE FRUIT WEIGHT 2017/2018 Crop Year [Internet]. 2018. Available from: https://www.almonds.com/sites/default/files/17-18_whole_crop_position_report_addendum .pdf [Accessed: 01 December 2021]

[141] Dehghani S, Nouri M, Baghi M. The effect of adding walnut green husk extract on antioxidant and antimicrobial properties of ketchup. Journal of Food and Bioprocess Engineering. 2019;**2**(2): 93-100

[142] Ramezani N, Raji F, Rezakazemi M, Younas M. Juglone extraction from walnut (*J. regia* L.) green husk by supercritical CO_2: Process optimization using Taguchi method. Journal of environmental. Chemical Engineering. 2020;**8**(3):103776

[143] Jahanban-Esfahlan A, Ostadrahimi A, Tabibiazar M, Amarowicz R. A comparative review on the extraction, antioxidant content and antioxidant potential of different parts of walnut (*J. regia* L.) fruit and tree. Molecules. 2019;**24**(11):2133

[144] Cosmulescu S, Trandafir I, Nour V. Seasonal variation of the main individual phenolics and juglone in walnut (*J. regia*) leaves. Pharmaceutical Biology. 2014;**52**(5):575-580

[145] Castillo AR, Silva-del-Río N, St-Pierre N, Weiss WP. Composition of diets fed to different groups of lactating cows on California dairies. Journal of Dairy Science. 2012;**95**(Suppl 2):360

[146] Çelik İ, Demirer GN. Biogas production from pistachio (*P. vera* L.) processing waste. Biocatalysis and Agricultural Biotechnology. 2015;**4**(4):767-772

[147] Holtman KM, Offeman RD, Franqui-Villanueva D, Bayati AK, Orts WJ. Countercurrent extraction of soluble sugars from almond hulls and assessment of the bioenergy potential. Journal of Agricultural and Food Chemistry. 2015;**63**(9):2490-2498

[148] Singh S, Dhanjal DS, Thotapalli S, Sharma P, Singh J. Importance and recent aspects of fungi-based food ingredients. In: Singh J, Gehlot P, editors. New and Future Developments in Microbial Biotechnology and Bioengineering. Amsterdam, Netherlands: Elsevier; 2020. pp. 245-254

[149] Wang K, Zhang R. Production of polyhydroxyalkanoates (PHA) by *Haloferax mediterranei* from food waste derived nutrients for biodegradable plastic applications. Journal of Microbiology and Biotechnology. 2021;**31**(2):338-347

[150] Barzee TJ, Cao L, Pan Z, Zhang R. Fungi for future foods. Journal of Future Foods. 2021;**1**(1):25-37

[151] Queirós CS, Cardoso S, Lourenço A, Ferreira J, Miranda I, Lourenço MJ, et al. Characterization of walnut, almond, and pine nut shells regarding chemical composition and extract composition. Biomass Conversion and Biorefinery. 2020;**10**(1):175-188

[152] Açıkalın K, Karaca F. Fixed-bed pyrolysis of walnut shell: Parameter

effects on yields and characterization of products. Journal of Analytical and Applied Pyrolysis. 2017;**125**: 234-242

[153] Taghizadeh-Alisaraei A, Assar HA, Ghobadian B, Motevali A. Potential of biofuel production from pistachio waste in Iran. Renewable and Sustainable Energy Reviews. 2017;**72**:510-522

[154] Harini K, Mohan CC, Ramya K, Karthikeyan S, Sukumar M. Effect of *Punica granatum* peel extracts on antimicrobial properties in walnut shell cellulose reinforced bio-thermoplastic starch films from cashew nut shells. Carbohydrate Polymers. 2018;**184**: 231-242

[155] Jafarzadeh S, Jafari SM, Salehabadi A, Nafchi AM, Kumar US, Khalil HA. Biodegradable green packaging with antimicrobial functions based on the bioactive compounds from tropical plants and their by-products. Trends in Food Science & Technology. 2020;**100**:262-277

Chapter 10

Edible Coating

Kofi Owusu-Akyaw Oduro

Abstract

Postharvest losses are rampant due to lack of proper storage conditions and handling of the fresh food products. The perishable nature of fruits and vegetables makes their shelf life limited due to some extrinsic factors such as some environmental conditions and preservation conditions as well as some intrinsic factors such as respiration rate, ethylene production and transpiration. Among the other postharvest technologies available, edible coatings seems to be one novel method which has been verified to have a positive and safe approach to extending the shelf life of products. This type of packaging is made from various natural resources like polysaccharide, protein and lipid materials. Edible packaging materials can be divided into two main groups including edible coatings and edible films. It has so many benefits such as serving as a moisture barrier, oxygen scavenger, ethylene scavenger, antimicrobial properties among others. Different methods of application of the edible coating on the food materials include; dipping, spraying, brushing, layer by layer among others. There have been several verifications of the positive impact of edible coatings/films on pome fruits, Citrus fruits, Stone fruits, tropical and exotic fruits, berries, melon, tomatoes and others.

Keywords: postharvest technology, edible coating/films, water loss, shelf life

1. Introduction

The global production of fruits keep increasing as a result of the rise in the population demand, elevation in the living quality standard and the increase in health awareness of fresh food products especially fruits and vegetables. This is because fruits and vegetables play vital roles in healthy nutrition due to their vitamins, minerals, antioxidant content among others. According to FAOSTAT [1], within about 10 years, the production of fruits which include drupes, berries, pome fruits, melons and tomatoes increased from 2,587,570 in 2007 to 34,622,004 metric tonnes in 2017. However food production has been reported by Alexandratos and Bruinsma [2] that it should be increased by 60% in 2050. Thus the increase in production is needed in parallel with the growth of the global population. However, postharvest losses which result in the degradation of quantity and quality of the fruits after harvest constitute a serious challenge.

Though these fruits have very high nutritional values, they are highly perishable due to their high moisture content and nutritional value leading to the development of undesirable characteristics as well as issues of food safety. These fresh food products are susceptible to dehydration, mechanical injury, environmental stress, pathological breakdown and enzymatic attacks which leads to some nutritional, functional and sensorial losses and production of off flavour and also posing a level of threat in terms of possessing a level of toxicity. There is a level of reduction of the edible quality of the food products due to biochemical changes, physiological ageing and microbial infections during storage and transportation.

Therefore the gas composition greatly affects the shelf life of the products. Extension of the supply time of fruits and vegetables besides preserving their quality would have economic profits [3]. In this regard, post-harvest practices aiming to maintain the physicochemical composition during storage must be adopted.

Fruits are either climacteric or unclimacteric. The latter cannot ripen once removed from the plant but the former can ripen after being picked and produce more ethylene which makes them more susceptible to spoilage. Thus to inhibit the rate of deterioration of these fruits, these is a need to alter the gaseous environment or control it. For instance making use of packaging materials with low water vapour and oxygen permeability to reduce respiration but not too low oxygenated environment which can lead to anaerobic respiration which can also produce off-flavours.

Although MAP and CA technologies can be regarded as the most effective methods with extensive and successful applications, they are quite expensive and chemical treatments on the other hand have potential levels of toxicity. Low temperature storage might also lead to chilling injury and heat treatment also leads to nutrient losses, decreased weight, flavour and vitamin losses [4]. One novel postharvest technology to circumvent these limitations is the use of edible coating which can control and inhibit the deteriorative changes as well as increasing the shelf life of the products. Edible coating/films is a good candidate to help solve the cases of postharvest losses since it has mechanical, thermal, antimicrobial and even antioxidant properties.

Edible coating or films are biopolymers that are hugely being investigated for the packaging and preservation of food. Edible packaging materials are a type of packaging that could be eaten and have the biodegradable ability also provides a barrier against moisture, gases and solute movement. Edible coatings are usually made from biodegradable materials such as Lipid-, Protein- or Polysaccharide-based materials. This packaging material is either used via a film or using coating. The latter is usually in liquid form whiles the former usually forming a thin layer around the food product. Edible coatings can be defined as a thin layer of edible and environmentally friendly materials that could be consumed and provide a barrier to gases, microbes and moisture to food products. Application of these films is simple, eco-friendly, highly safe and low priced which makes it promising for preserving food products.

There has been several research works on the impact of edible coating on the physiological and microbial stability of some fresh produce. For instance, Li et al. [5] verifies that application of Cinnamaldehyde as an edible coating on banana showed a significant decrease in the weight loss and ripening rate of the banana. Also, application of protein isolate with organo-clay MMT on minimally processed papaya sliced also demonstrated a lower microbial growth and lower mass loss [6]. An increasing interest in edible films/coatings is an outcome of growing consumer awareness on healthy foods, and also due to negative impacts of non-biodegradable synthetic packaging materials on the environment.

Edible coatings/films helps to improve the appearance of horticultural produce by giving shine, hiding scars, suppressing decay and physiological disorder developments [7]. Edible coatings can be generally classified into three main groups; Protein-based edible coatings, Polysaccharide-based coatings and lipid-based coatings. The choice of active agents depends on the characteristics of the product and the type of polymeric matrix in the coatings. Active or functional compounds; Antioxidants, antimicrobials, nutrients, vitamins, anti-browning agents, enzymes and probiotics that could be applied into coating matrix to help preserving products quality.

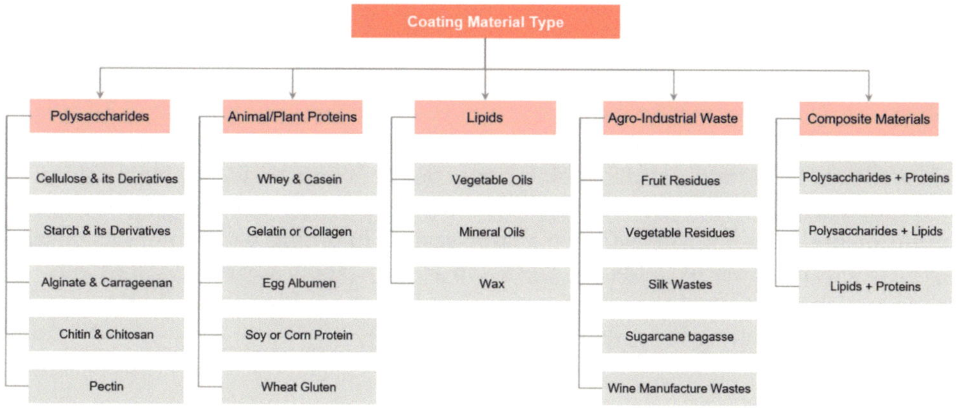

Source: FutureBridge Analysis

2. Classes of edible coating

2.1 Coatings based on Polysaccahrides

Polysaccharides are natural polymers used extensively to produce edible coating or films. Examples of polysaccharides used in the production of these films include; Pectin, cellulose, starch, chitosan, alginates and pullulan. Polysaccharides are the basic coatings that are considered to be an effective blocker of oxygen because of its ordered structure including a hydrogen network. However, polysaccharides form a poor barrier against water vapour because of its hydrophilic nature. They are usually used to improve the shelf life of meat products, vegetables, fruitc.

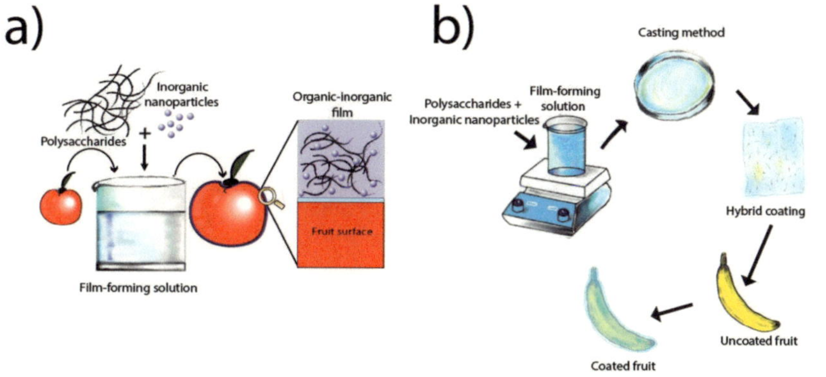

2.1.1 Starch

Starch is a polysaccharide that is composed of two different molecules which are amylose which is a linear polymer and amylopectin which is a highly branched polymer. Starch is widely used in coatings for food materials since it is abundant in nature and has a low cost. Several studies have been carried out to improve physicochemical and optical properties of starch-based edible films using Aloe Vera. Coatings based on starch are odourless and colourless. They possess less oxygen permeability and have an oil-free appearance. They can make an important contribution to decrease in the respiration rate for the fresh fruits and vegetables.

2.1.2 Chitosan

Natural chitin can go through a process called deacetylation to form Chitosan which is a polysaccharide that can be used in edible coating of food materials [8]. Chitosan is mostly used in coating materials for fruits and vegetables because of its antioxidative and antimicrobial properties. It is non-toxic, biodegradable, biocompatible and microbe-resistant, chitosan is currently attracting considerable attention and its scientific testing at a large scale is in progress to explore its possible applications in different fields [9]. Chitosan are partial permeable coatings and films, which can control the interior structure by diminishing transpiration rates and retarding ripening in foods and vegetables [10].

2.1.3 Alginate

Alginate is an unbranched polysaccharide and is composed of sodium salt of alginic acid that is derived from some species of brown algae. Alginates are indigestible natural polysaccharides acquired from seaweed and have been reported to be a stabilising and thickening in the food market. It has good film forming properties as it can form gels through crosslinking with divalent cations like Ca2+. For this reason, alginate finds interesting application for coating fresh and processed food items [11].

2.1.4 Gellan gum

Gellan gum consists of repetitive units of tetrasaccharides, and it is a well-recognised biopolymer due to its functional properties, eg) good hardness, high transparency, smooth surfaces and reduced water vapour permeability.

2.2 Pullulan-based coatings

Pullulan is a polysaccharide which is usually a thickener that may form effective films. The use of pullulan edible films and coatings in combination with chitooligosaccharide which has antibacterial properties and glutathione which is also a powerful reducing agent. This makes it effective in increasing the shelf life of various food products.

2.2.1 Cellulose

Cellulose is also a linear chain polysaccharide which is a major component of plant cell wall which has a large number of intra-molecular hydrogen bonds causing its water insolubility with highly associated crystalline structure [12]. The native cellulose has very low water solubility properties and is a less suitable film forming material. However, various chemically modified forms of cellulose like carboxymethyl cellulose, methylcellulose, hydroxypropyl cellulose and hydroxypropylmethyl cellulose are quite suitable for film and coating applications.

2.2.2 Carboxymethyl cellulose

An anionic linear and long chain compound that consists of glucopyronosyl units with high molecular weight providing strength and structural integrity in edible coatings. They exhibit excellent oxygen, aroma, and oil barrier and anti-senescence properties.

2.2.3 Pectin

Pectin, main compound of plant cell walls found in middle lamella of plant cells. They are complex heteropolymers made up of D-galacturonic acid units that may present variations in composition, structure and molecular weight [13].

2.3 Protein-based coatings

Proteins generally occur in the form of globular proteins or fibrous proteins. Fibrous proteins are insoluble in water and generally play the role of a basic structural element of animal tissues, they are also soluble in aqueous solutions of salt, bases or acids and perform different activities in living systems. Various types of globular proteins such as corn zein, whey protein, wheat gluten and soy protein are involved in edible coatings/films. A dispersion or protein solution is taken into consideration to create coatings and films, and the solvent that is taken into consideration for playing this role is generally restricted to ethanol water combinations, or simply water or ethanol.

Protein-based coatings which include the use of casein, gluten and soy protein serve as good oxygen blockers and thus help preserve the food products from any deteriorative reactions. Proteins are reported to impart good mechanical properties and gas barrier properties.

2.4 Corn zein-based films and coatings

Corn is a major source of zein which is a prolamin protein that can be dissolved in 70–80% ethanol and hydrophobic in nature. Edible coating made from zein shows very good film properties. They are good moisture blockers than other films.

2.5 Gelatin-based coatings and films

Gelatin is a hydrophobic protein usually found in wheat which is also a globular protein and also used in some edible coatings/films due to its low cost and availability. Gelatin coatings usually depict good transparency, mechanical and barrier properties and can be manufactured via an extrusion or casting process. The nature of the gluten has significant impact on its filming properties.

2.6 Lipid-based coatings and films

Lipids are naturally hydrophobic in nature making them very good materials to be used in edible coating since they can help resolve moisture migration into the fresh food product which can cause some significant deteriorative changes in the food material. Some example of lipids used in edible coatings include wax and paraffin [14].

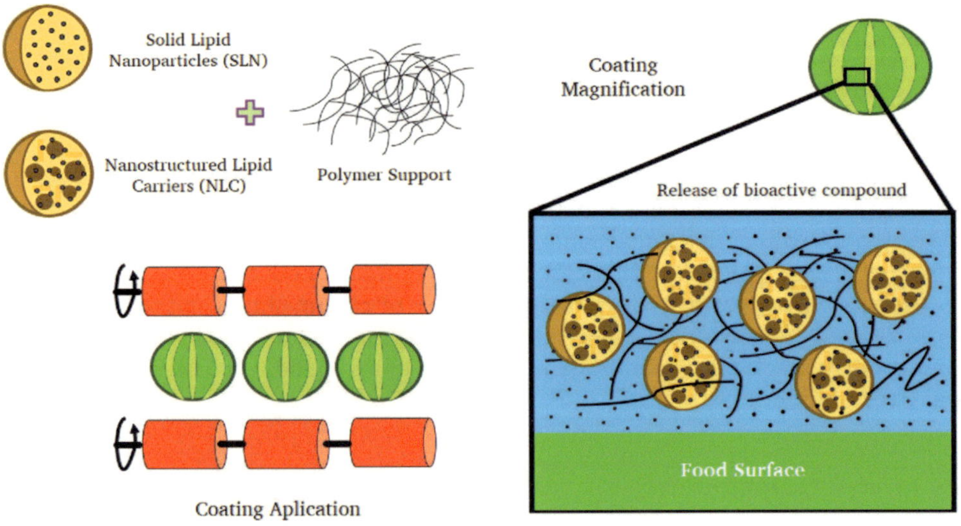

3. Methods of application of edible coating/films

3.1 Dipping

This technique is the oldest commercial technique but still relevant until now. The concept of dipping technique is by immersing the fresh food produce into the coating solution to allow complete wetting of the surface of the food material. After that the coating solution is drained out to remove excess coating from the food surface. Finally the fruit is dried to form a well intact coating with the food surface. This can be applied to a wide range of viscous coating solutions.

3.2 Layer by layer method

Layer by layer method is based on alternate deposition of oppositely charged polyelectrolytes that result in a more effective control of the coating properties and functionality. This method leads to the production of several layers of the films which can help to improve the effectiveness of the edible coating.

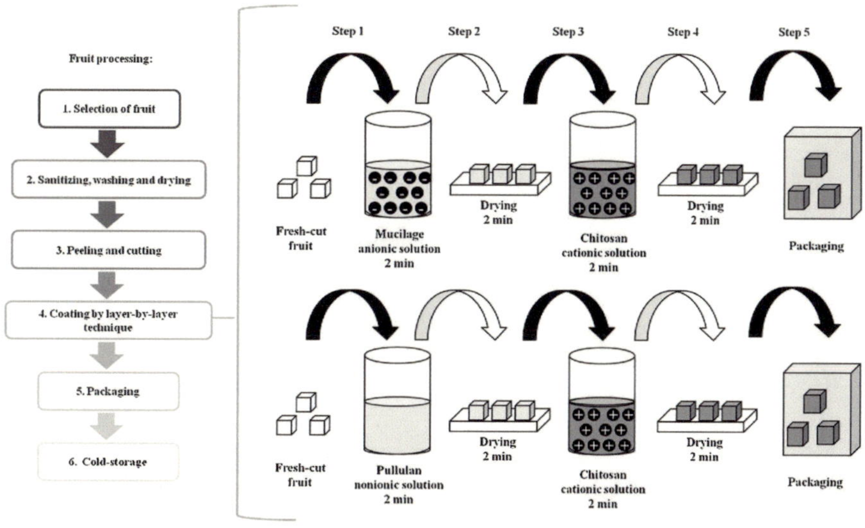

3.3 Vacuum impregnation technique

Vacuum impregnation technique is a further advancement of the dipping method. The difference is having a vacuum environment during fruit dipping. That is, instead of dipping the food material in a normal dipping tank, the fresh food is submerged in an airtight vacuum application. The food material is subjected to atmospheric restoration while it remains immersed in the coating solution under atmospheric pressure.

3.4 Spraying method

Spraying method is more suitable for less viscous coating solutions which can be sprayed at high pressure. Formation of polymeric coating using spraying system is affected by drying time and temperature. The advantage of applying the spraying technique is, the surface area of the liquid coating increase through the formation of droplets and distribution over the fruit surface.

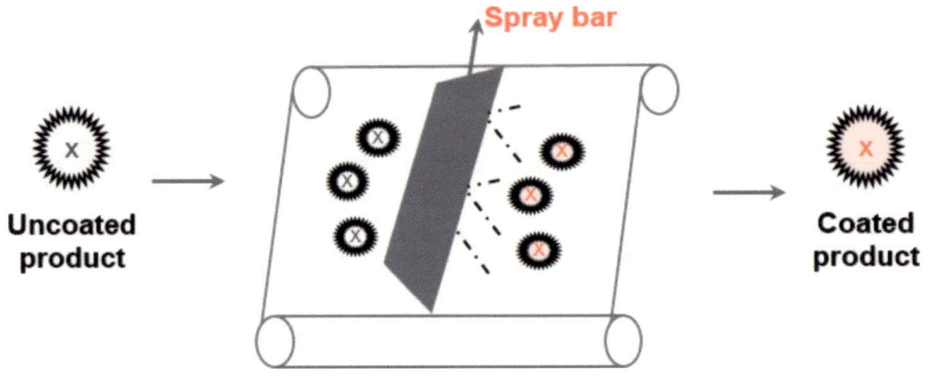

3.5 Foaming and dripping method

Foaming and dripping method are considered as traditional methods in coating application. These methods are now gaining low popularity among researchers and industrial practitioners in fruit industries. With the dripping technique, the coating is being applied directly to the fruit surface using brushes However with the foam application, a foaming agent is added to the coating. Then, compressed air is blown into the air of applicator tank. Extensive tumbling action is applied to break the foam for uniform distribution.

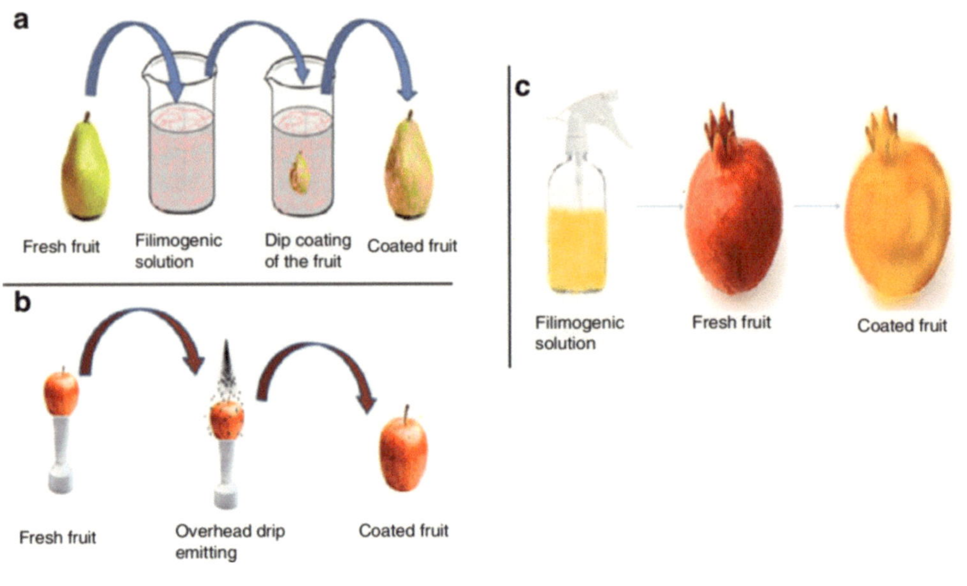

4. Benefits of edible coatings/films

4.1 Moisture barrier

These films prevent moisture loss, aroma loss or water uptake by the food material or even penetration of oxygen which produces a good storability condition for these food products, Edible coating enhance the texture and improves the product appearance and prolong the shelf life by creating semi-permeable barriers. Emamifar and Bavaisi [15] developed a bio-nanocomposite coating with sodium alginate and nano-ZnO and applied it on strawberry. The results revealed a significant weight retention than those without the films. Again Titanium and silver nanocomposite packaging displayed same results on mangoes [16].

4.2 Oxygen scavengers

The presence of oxygen can have considerable detrimental effects on some packaged fresh food products. Some edible films have been found to contain some oxygen scavengers and humidity control systems which play an important role in reducing gases contributing to the spoilage of fruits and vegetables. Resende et al. [17] indicated that the coating of chitosan/cellulose nonofibril minimises the oxygen diffusion, decreases respiration and delays strawberry oxidation by ascorbic acid reaction.

4.3 Ethylene scavenger

Ethylene control in storage time plays a significant role in extending the shelf life of the fresh produce. Kaewklin et al. [18] determined the ethylene control activity of chitosan-TiO2 nanocomposites on tomato showed lower levels of ethylene concentration.

4.4 Antimicrobial properties

One of the main contamination reasons for fruit and vegetable is the lack of proper packaging. An antimicrobial active packaging system loaded with antimicrobial agents can be applied to minimise the spoilage of fresh produce to control their microbial growth. Some studies also proved this as strawberries coated with 1.5% sodium alginate and nano ZnO showed the lowest growth of micro-organisms. The Antimicrobials in the edible coatings enhance the shelf life and safety of fruits and vegetables by preventing microbial growth and damages [19]. Some of the antimicrobial substances include organic acids such as citric acid and lactic acid, Microbial bacteriocins like Lactic acid bacteria and some polypeptides such as lysozymes [20].

4.5 Antibrowning and antioxidant properties

Enzymatic browning in minimally-processed fruits and vegetables is linked to discoloration and discoloration of phenolic compounds catalysed by polyphenol oxidase (PPO) enzyme, which converts polyphenolic substrates to dark pigments in the presence of oxygen. Edible coating especially incorporated with antibrowning substances can control PPO activity, and in the other hand, can provide a strong barrier for oxygen. The antibrowning substances mostly used are ascorbic acid, thiol-containing compounds (cysteine and glutathione), carboxylic acids (citric and oxalic acid), phenolic acids and resorcinols. These reduce o-quinones resulted from the action of PPO enzymes, back to their phenolic substrates [21].

5. Texture modifiers for inhibition of physical damages

Pectolytic enzymes leads to the loss of firmness in fruit tissues and so any attempt to inhibit this enzyme's activity will result in firmness retention. Application of edible coatings containing active substances called texture enhancers could minimise the textural softening of fruits and vegetables during storage. These compounds retard plygalacturonase activity and preserve structural integrity of membrane. To control softening phenomena in fresh-cut fruits calcium salts are commonly used and considered as firmness retainers.

6. Nutraceuticals for preservation of nutritional quality

Nutraceuticals enhance the nutritional profile of low-micronutrient products; Minerals, vitamins and bioactive compounds are potential Nutraceuticals compounds that can be incorporated in formulation of active coatings to enhance the nutritional value of some fruits and vegetables, where these micronutrients are present in low quantities [22].

7. Application of edible coating/films on some selected food products

7.1 Apple

Apple which is a Pome fruit has undergone various research studies which proves the effectiveness of edible coating in the preservation of this fruit. For instance a research finding by Guerreiro et al. [23] showed a significant reduction of microbial load on the food product and resulted in a prolonged shelf life.

7.2 Citrus

FAOSTAT [1] reported that citrus is one of the main crops in the world with a total production of 18.9 million tonnes in 2017. Similar to the other fresh produce, postharvest losses are the major problem in the citrus production chain. Arnon et al. [24] developed a by-layer polysaccharides-based edible coating for mandarins using CMC as the internal layer whiles the chitosan was used as the external layer. The result demonstrated that the quality of the citrus fruits such as the gradient of the glossiness and peel colour were evenly improved.

7.3 Mango

Mango which is also a drupe fruit along with cherries and peaches have shown some significant improvement in terms of its shelf life upon the addition of edible coating/film materials n it. Though Mango is most preferred due to its appealine organoleptic properties it has been shown to undergo rapid deterioration after harvest.

Paladines et al. [25] investigated the impact of roseship oil with aloe vera gel von deferring ripening and preserving the postharvest quality of a number of stone fruits. The results indicated that the aloe Vera coating inhibited the formation of ethylene, decreased the respiration rate and delayed the changes the fruit colour and firmness. Again studies using guar gum and ginseng extract on sweet cherry showed a significant delay in the production of malondialdehyde [26].

7.4 Berries

Berry fruits such as blackberry are commonly used in the human diet either fresh or in processed form. Berries are small fruits that contain high antioxidant benefits. Several studies have been on the integrity of edible nano-coatings of curcumin and limonene liposomes integrated with methyl cellulose and its impact on the quality of strawberries and this showed the coating was found to be effective in regulating fungal decay in strawberries [27].

7.5 Melon

Carvalho et al. [28] stated that most of the cultivated melons are eaten as value added particularly fruits, especially for fresh-cut products. Though these food products have been found to deteriorate quite easily due to various biochemical processes, a lot of research has proven the effectiveness of edible coating in inhibiting the deteriorative changes [29].

7.6 Tomatoes

Tomatoes are one of the most vulnerable food products in the world due to their delicate structure. This obviously makes storage of these food products quite difficult since they even undergo rapid deteriorative changes after harvesting. There have been some successful findings on the positive impact of edible coatings on the shelf life of Tomato (**Table 1**).

Coating material	Food product	Impact on product	References
Alginate and chitosan	Guava	Improved shelf life	Arroyo *et al.* [30]; S. Panahirad et al. [31]
Glycerol and carnauba wax with aloe vera	Mango	•Delayed loss of firmness and weight •Less changes in colour, pH and Brix value •Controlled rate of respiration	Peres et al. [32]; Maan et al. [33]
Pectin	Tomatoes	• Weight loss retention • Delay in ripening index	Abebe *et al.* [34]; B. Manringgal et al. [35]
Carboxyl methylcellulose	Avocado	•Firmness and weight loss retention •Reduce the respiration rate •Antimicrobial •Increase the shelf life	Tesfay *et al.* [36]; Manringgal *et al.* [35]
Gelatin, Guar, Chitosan	Barhi date	Extended the shelf life of Barhi date fruits in comparison with the control sample	Abu-Shama *et al.* [37]; N.A. Al-Tayyar *et al.* [38]
Beewax, Chitosan	Strawberries	•Prevention of fungal infection, reduced weight loss and respiration rate	Velickova *et al.* [39]; N.A. Al-Tayyar *et al.* [38]

Table 1.
Impact of edible coating materials on food products.

Application of edible coatings have demonstrated a positive result in terms of improving the shelf life and preserving the quality of tropical fruit. Edible coatings have been added to pitaya [40], soursop [41], pineapple [42], papaya [43], banana [44], longan [45], and guava [46].

8. Conclusion

Edible coating is a very interesting field of study that could revolutionise the postharvest industry as we know it. These materials are biodegradable, eco-friendly and has less to no negative impact on the food product. There has been so many proven evidences on the positive impact of edible coatings and films on some food products.

Author details

Kofi Owusu-Akyaw Oduro
Kwame Nkrumah University of Science and Technology (Knust), Kumasi, Ghana

*Address all correspondence to: okofi4915@gmail.com

References

[1] FAOSTAT (2019). Food and agriculture organization corporate statistical database 2019. Available from: http://www.fao.org/faostat/en/#data (Accessed: August 2021)

[2] Alexandratos N, Bruinsma J. World agriculture towards 2030/2050: The 2012 Revision. Rome: FAO ESA working paper No 12-03; 2012

[3] Bilal Hassan, Shahzad Ali Shahid Chatha*, Abdullah Ijaz Hussain, Khalid Mahmood Zia, Naseem Akhtar (2018a). Recent advances on polysaccharides, lipids and protein based edible films and coatings: A review. International Journal of Biological Macromolecules. 109: 1095 – 1107.

[4] Beigmohammadi F, Peighambardoust SH, Hesari J, Azadmard-Damirchi S, Peighambardoust SJ, Khosrowshahi NK. Antibacterial properties of LDPE nanocomposite films in packaging of UF cheese. Lebensmittel-Wissenschaft und -Technologie-Food Science and Technology. 2016;**65**:106-111. DOI: 10.1016/j.lwt.2015.07.059

[5] Li J, Sun Q, Sun Y, Chen B, Wu X, Le T. Improvement of banana postharvest quality using a novel soybean protein isolate/ cinnamaldehyde/zinc oxide bionanocomposite coating strategy. Sci Hortic (Amsterdam). 2019;**258**:108786

[6] Cortez-Vega WR, Pizato S, de Souza JTA, Prentice C. Using edible coatings from Whitemouth croaker (Micropogonias furnieri) protein isolate and organo-clay nanocomposite for improve the conservation properties of fresh-cut 'Formosa'papaya. Innovative Food Science and Emerging Technologies. 2014;**22**:197-202

[7] Ncama K, Magwaza LS, Mditshwa A, Tesfay SZ. Plant-based edible coatings for managing postharvest quality of fresh horticultural produce: A review. Food Packaging and Shelf Life. 2018;**16**:157-167. DOI: 10.1016/j.fpsl.2018.03.011

[8] Khatri D, Panigrahi J, Prajapati A, Bariya H. Attributes of *Aloe vera* gel and chitosan treatments on the quality and biochemical traits of post-harvest tomatoes. Scientia Horticulturae. 2020;**259**:108837. DOI: 10.1016/j. scienta.2019.108837

[9] Du Y, Zhao Y, Dai S, Yang B. Preparation of water – soluble chitosan from shrimp shell and its anti-bacterial activity. Innovative Food Science and Emerging Technologies. 2009;**10**(1): 103-107

[10] Qi H, Hu W, Jiang A, Tian M, Li Y. Extending shelf-life of Fresh-cut 'Fuji' applies with chitosan-coatings. Innovative Food Science and Emerging Technologies. 2011;**12**(1):62-66

[11] Carvalho RL, Cabral MF, Germano TA, Carvalho WMD, Brasil IM, Gallão M, et al. Nanocomposite coating based on sodiumalginate and nano-ZnO for extending the storage life of fresh strawberries (Fragaria×ananassa Duch.). J Food Meas Charact. 2020;**14**:1012-1024. DOI: 10.1007/s11694-019-00350-x

[12] Suput DZ, Lazic VL, Popovic SZ, Hromis NM. Edible films and coatings: Sources, properties and application. Food and Feed Research. 2015;**42**(1): 11-22. DOI: 10.5937/FFR1501011S

[13] Lara-Espinoza C, Carvajal-Milían E, Balandŕan-Quintana R, López-Franco Y, Rasćon-Chu A. Pectin and pectin-based composite materials: Beyond food texture. Molecules. 2018;*23*(4):942. DOI: 10.3390/molecules23040942

[14] Yousuf B, Wu S, Siddiqui MW. Incorporating essential oils or compounds derived into edible coatings:

Effect on quality and shelf life of fresh/fresh-cut produce. Trends in Food Science and Technology. 2021;**108**:245-257

[15] Emamifar A, Bavaisi S. Nanocomposite coating based on sodium alginate and nano ZnO for extending the storage life of fresh strawberries (Fragaria × ananassa Duch.). J Food Meas Charact. 2020;**14**:1012-1024. DOI: 10.1007/s11694-019- 00350-x

[16] Chi H, Song S, Luo M, Zhang C, Li W, Li L, et al. Effect of PLA nanocomposite films containing bergamot essential oil, TiO2 nanoparticles, and Ag nanoparticles on shelf life of mangoes. Sci Hortic (Amsterdam). 2019;**249**:192-198

[17] Resende NS, Gonçalves GAS, Reis KC, Tonoli GHD, Boas EVBV. Chitosan/cellulose nanofibril nanocomposite and its effect on quality of coated strawberries. Journal of Food Quality. 2018;**2018**. DOI: 10.1155/2018/1727426

[18] Kaewklin P, Siripatrawan U, Suwanagul A, Lee YS. Active packaging from chitosantitanium dioxide nanocomposite film for prolonging storage life of tomato fruit. International Journal of Biological Macromolecules. 2018;**112**:523-529

[19] Jafarzadeh S, Nafchi AM, Salehabadi A, Oladzad-abbasabadia N, Jafari SM. Application of bio-nanocompositie films and edible coatings for extending the shelf life of fresh fruits and vegetables. Advances in Colloid and Interface Science. 2021;**116**:218-231

[20] Salas-Méndez EDJ, Vicente A, Pinheiro AC, Ballesteros LF, Silva P, Rodríguez-García R. Application of edible nanolaminate coatings with antimicrobial extract of Flourensia cernua to extend the shelf-life of tomato (Solanum lycopersicum L.) fruit. Postharvest Biology and Technology. 2019;**150**:19-27

[21] El-Hosry L, Auezova L, Sakr A, Hajj-Moussa E. Browning susceptibility of white wine and antioxidant effect of glutathione. International Journal of Food Science and Technology. 2009;**44**(12):2459-2463

[22] Basaglia RR, Pizato S, Santiago NG, de Almeida MMM, Pinedo RA, Cortez-Vega WR. Effect of edible chitosan and cinnamon essential oil coatings on the shelf life of minimally processed pineapple (*Smooth cayenne*). Food Bioscience. 2021;**41**:67-86

[23] Guerreiro AC, Gago CML, Faleiro ML, Miguel MGC, Antunes MDC. The effect of edible coating on the nutrional quality of Bravo de Emolfe freshcut apple through shelf life. LWT-Food Science and Technology. 2017;**75**:210-219

[24] Arnon H, Granit R, Porat R, Poverenov E. Development of polysaccharides based edible coatings for citrus fruits: A layer-by-layer approach. Food Chemistry. 2015;**166**:465-472

[25] Paladines D, Valero D, Valverde JM, Díaz-Mula H, Serrano M, Martínez-Romero D. The addition of rosehip oil improves the beneficial effect of Aloe vera gel on delaying ripening and maintaining postharvest quality of several stone fruit. Postharvest Biology and Technology. 2014;**92**:23-28

[26] Rivaa SC, Oparaab UO, Fawole OA. Recent developments on postharvest application of edible coatings on stone fruit: A review. Scientia Horticulturae. 2020;**262**:45-56

[27] Nora SM, Dinga P. Trends and advances in edible biopolymer coating for tropical fruit: A review. Food Research International. 2020;**134**:109208

[28] Carvalho RL et al. Chitosan coating with trans-cinnamaldehyde improves structural Integrity and antioxidant metabolism of fresh-cut melon.

Postharvest Biology and Technology. 2016;**113**:29-39

[29] Poverenov E, Danino S, Horev B, Granit R, Vinokur Y, Rodov V. Layer by-layer electrostatic deposition of edible coating on fresh cut melon model: Anticipated and unexpected effects of alginate-chitosan combination. Food and Bioprocess Technology. 2014;**7**:1424-1432

[30] Arroyo, B. J., Bezerra, A. C., Oliveira, L. L., Arroyo, S. J., Melo, E. A. de, & Santos, A. M. P. (2020). Antimicrobial active edible coating of alginate and chitosan add ZnO nanoparticles applied in guavas (Psidium guajava L.). Food Chemistry, 309, 125566

[31] Panahirad S, Dadpour M, Peighambardoust SH, Soltanzadeh M, Gullon B, Alirezalud K, et al. Applications of carboxymethyl cellulose and pectin-based active edible coatings in preservation of fruits and vegetables: A review. Trends in Food Science and Technology. 2021, 110;**663**:-673

[32] Perez AFT, TID A, FJI R. Conservation of minimally processed mango tommy atkins by applying an aloe vera (Aloe barbandensis miller) coating. Vitae. 2016;**23**:65-77

[33] Maan AA, Ahmed ZFR, Khan MKI, Riaz A, Nazir A. Alow vera gel, an excellent base material for edible films and coatings. Trends in Food Science and Technology. 2021;**116**:329-341

[34] Abebe Z, Tola YB, Mohammed A. Effects of edible coating materials and stages of maturity at harvest on storage life and quality of tomato (Lycopersicon Esculentum Mill.) fruits. African Journal of Agricultural Research. 2017;**12**:550-565

[35] Maringgala B, Norhashila Hashima D, Tawakkalb ISMA, Mohamed MTM. Recent advance in edible coating and its effect on fresh/fresh-cut fruits quality. Trends in Food Science and Technology. 2020;**96**:253-267

[36] Tesfay SZ, Magwaza LS, Mbili N, Mditshwa A. Carboxyl methylcellulose (CMC) containing moringa plant extracts as new postharvest organic edible coating for Avocado (Persea americana Mill.) fruit. Scientia Horticulturae. 2017;**226**:201-207

[37] Abu-Shama HS, Abou-Zaid FOF, El-Sayed EZ. Effect of using edible coatings on fruit quality of Barhi date cultivar. Scientia Horticulturae. 2020;**265**:109262

[38] Al-Tayyar NA, Youssef AM, Al-Hindi RR. Edible coatings and antimicrobial nanoemulsins for enhancing shelf life and reducing foodborne pathogens of fruit and vegetables: A review. Sustainable Materials and Technologies. 2020;**26**:45-50

[39] Velickova E, Winkelhausen E, Kuzmanova S, Alves VD, Moldão-Martins M. Impact of chitosan-beeswax edible coatings on the quality of fresh strawberries (Fragaria ananassa cv Camarosa) under commercial storage conditions. LWT - Food Science and Technology. 2013;**52**(2):80-92

[40] Fan P, Huber DJ, Su Z, Hu M, Gao Z, Li M, et al. Effect of postharvest spray of apple polyphenols on the quality of fresh-cut red pitaya fruit during shelf life. Food Chemistry. 2018;**243**:19-25

[41] Moreno-Hernández CL, Sáyago-Ayerdi SG, García-Galindo HS, Montes De Oca MM, Montalvo-González E. Effect of the application of 1-Methylcyclopropene and wax emulsions on proximate analysis and some antioxidants of soursop (Annona muricata L.). The Scientific World Journal. 2014;**1-7**: 896853

[42] Azarakhsh N, Osman A, Ghazali HM, Tan CP, Adzahan NM. Lemongrass essential oil incorporated into alginate-based edible coating for shelf-life extension and quality retention of fresh-cut pineapple. Postharvest Biology and Technology. 2014;**88**:1-7

[43] Mendy TK, Misran A, Mahmud TMM, Ismail SI. Application of Aloe vera coating delays ripening and extend the shelf life of papaya fruit. Scientia Horticulturae. 2019;**246**:769-776

[44] Alali AA, Awad MA, Al-Qurashi AD, Mohamed SA. Postharvest gum Arabic and salicylic acid dipping affect quality and biochemical changes of 'Grand Nain' bananas during shelf life. Scientia Horticulturae. 2018;**237**:51-58

[45] Lin MG, Lasekan OL, Saari N, Khairunniza-Bejo S. The effect of the of edible coatings on or before ultraviolet treatment on postharvested longan fruits. Journal of Food Quality ID. 2017:5454263

[46] Silva WB, Silva GMC, Santana DB, Salvador AR, Medeiros DB, Belghith I. Chitosan delays ripening and ROS production in guava (Psidium guajava L.) fruit. Food Chemistry. 2018;**242**: 232-238

Chapter 11

Postharvest Processing, Value Addition and Marketing of Mushrooms

Mahesh Prasad Thakur, Harvinder K. Singh and Chandra Shekhar Shukla

Abstract

Mushrooms are macrofungi having a higher content of water (80–90%) and multinutrients. The presence of various phytochemicals, enzymes, primary metabolites and secondary mycometabolites results in poor shelf-life, quick deterioration, and huge postharvest losses (30–35%). Fresh mushrooms are short lived (1–8 days). Value chain management is thus necessary from the production to its harvest to meet the food and nutritional requirements. Every effort was made to extend the shelf-life of mushrooms for either short period or long period of storage. Washing or pretreatment, packaging, transport and marketing were some of the important standardized techniques for short-term storage of mushroom. On the other hand, drying, pickling and steeping preservation methods were some other techniques to extend the shelf-life of mushroom for a longer period of time during storage. Value addition of mushroom enhanced the quality and addressed the demand for ready-made or ready-to-make food products. Fresh/dry oyster mushroom in various proportions (5–10%) was used to prepare mushroom paratha, mushroom suji, mushroom sandwich, mushroom chakli, mushroom seb, mushroom-based biofortified wheat flour, mushroom-based papad, nuggets, mushroom bijoura, biscuits, etc. Several mushroom-based, value-added products like Royal Oyster Capsules were prepared by Self Help Groups women at Kapadah (Kabirdham).

Keywords: post harvest, mushroom processing, preservation, value addition, shelf life *Pleurotus flabellatus*, sporophores, steeping solutions

1. Introduction

Mushrooms are highly perishable in nature and subjected to change in ways that make them unacceptable for human consumption. It has high water content (85–95%), which is lost rapidly by evaporation and transpiration making mushroom discolored, disfigured and unfit for consumption. The rate of water loss depends on mushroom structures, relative humidity, temperature, air movement and atmospheric pressure during storage. Mushrooms represent one of the most perishable commodities, being so delicate by nature and hence need special postharvest treatments. A number of physiological processes take place in freshly cut mushrooms when it is stored (pileus and veil opening, stipe elongation, browning, etc.) resulting in maturation, senesce, and decrease in commercial and nutritional values

[1, 2]. Burton et al. [3] found varying changes in color, size, clearness, firmness, maturity stage, blemish-free, flavor, nutritional value and safety by pre-harvest treatments, postharvest processing and storage conditions. The major limitations in mushroom marketing include wilting and shriveling due to rapid water loss, which render them unfit for marketing and consumption [4]. The shelf life of mushroom can often be extended by pretreatments and/or storage at chilling temperatures (above freezing and below 0°C), chemical preservation, and drying processes. Adsule et al. [5] reported the preservation of *Pleurotus sajor-caju* fruit bodies in steeping solution containing 5% salt, 0.20% citric acid and 0.1% potassium metabisulphite (KMS) up to three months without losing much of its organoleptic quality. *Pleurotus sajor-caju* and *Volvariella volvacea* (24 days) fruit bodies were successfully stored without spoilage at room temperature in chemical solutions consisted of salt, sugars, acids and preservatives [6, 7]. Water blanched white button mushroom can be successfully stored from 2 to 5 weeks to 3 months by preserving in steeping solutions of various concentrations of salt, sugars, acids and preservatives [8, 9]. Gormley and O' Riordain [10] reported the storage of *Pleurotus ostreatus* fruit bodies for 3 months at −30°C. Storage of *Pleurotus flabellatus*, *P. sapidus* and *P. ostreatus* fruit bodies up to 15 days in different thickness of polyethylene films were reported by several workers [11–13]. Spoilage of mushrooms during storage was associated with the presence of microorganisms dominated by bacteria, fungi and enzymes which strongly influenced the physiology and shelf life. Mushrooms need to be properly processed in order to extend their shelf life so that they can be used during off-season and also add value to the product. Value addition of mushroom with several traditional recipes can be achieved by adopting appropriate postharvest technology to process surplus mushrooms into various value-added products (soup powder, pickles, chips, paste and ketchup, pâté, noodles and pasta, biscuits, and nuggets), mushroom-based flavor enhancers or as additives in beverages and beauty products [14, 15]. The value-added products are the urgent need for the mushroom growers not only to reduce the losses, but also to enhance the income by value-addition and boosting mushroom consumption [16]. Biofortification or value addition of mushroom in present days is becoming very common to enhance quality, shelf life, alleviate under or malnutrition and reaching among various sections of the society [15]. Keeping these in view, efforts were made to enhance the shelf life of button mushrooms and its varieties by dipping them in the solutions of ethylene diamine tetra-acetic acid (EDTA) at different concentrations. Effect of packaging materials (thickness) on shelf life of button mushrooms and its varieties with respect to the quality parameters under both ambient/refrigerated conditions was studied. Similarly, studies were conducted on methods of drying of oyster mushroom (*Pleurotus florida*) on weight loss and other quality parameters. It was attempted to examine the shelf life and quality parameters of *P. flabellatus* in further dilutions and different combinations of steeping solutions. Efforts on processing and value addition was also attempted to extend the shelf life and prepare different mushroom based value added products.

2. Enhancement of shelf life of button mushroom

Effect of dipping treatment on the quality and shelf life of the button mushroom (*Agaricus bisporus*) was studied at Mushroom Research Laboratory (AICRP on Mushroom) of the Department of Plant Pathology, Indira Gandhi Krishi Vishwavidyalaya (IGKV), Raipur for two years. The fresh fruiting body of four varieties of button mushroom, S 649, SSI 4035, Pant 31 and NCS 100 of size 25 ± 2 cm or available size fruit bodies were washed in solutions containing EDTA at 75, 125, 200 ppm & 0.05% of potassium meta bisulfate (KMS) concentration. Thereafter, the fruit bodies were taken out, put in blotter paper to remove moisture and stored at ambient temperature and refrigerated temperature for different durations after packaging in 100 gauge polypropylene bags.

2.1 Shelf life extension of button mushroom varieties/strains

Effect of dipping treatment in different concentrations of EDTA solutions and KMS was studied on weight loss and other quality parameters and shelf life of four varieties/strains of *A. bisporus viz.*, S 649, SSI 4035, Pant 31 and NCS 100.

Loss in weight of fruit bodies of *A. bisporus* in cv. S 649 was considerably more under both ambient temperature and refrigerated conditions when it was kept in EDTA between 75 and 125 ppm concentrations. However, it was considerably less when the fruit bodies were subjected to EDTA 500 and 200 ppm under both the conditions. Fruit bodies under refrigerated conditions could be well preserved for longer period of time (11–13 days) and loses less weight (0.58 g) without being much influence on whiteness, texture and opening of the gills. However, there was considerable reduction in weight of fruit bodies (0.96 g), period of storage (3.50–4.50 days) and other quality parameters when the fruiting bodies were preserved under ambient temperature conditions (**Table 1**). Short shelf-life (1–3 days) of mushrooms at ambient temperatures (ca. 22°C) was reported by Burton and Twyning [17] and Wakchaure [18] while longer shelf-life (for about 7–9 days) was reported by Gormley [19] at lower or refrigerated temperature (0–1°C) which are in agreement with the present findings.

The fruiting bodies of *A. bisporus* cv. NCS 100 could be well preserved for little longer period of time (13–16 days) and loses less weight (0.64 g) under refrigerated conditions without being much influence on whiteness, texture and opening of the gills. On the other hand, there was considerable reduction in weight of fruit bodies (0.99 g) and period of storage (3–5 days) influencing whiteness, texture and gill opening under ambient temperature conditions (**Table 2**). Under ambient temperature conditions, the rate of water loss was more resulting in shorter shelf life and change of mushroom structures which are also influenced by relative humidity, temperature, air movement and atmospheric pressure of the storage environment. It ultimately affect the shelf life, quality and had adverse effect on mushroom marketing mainly due to wilting and shriveling rendering mushrooms unfit for marketing and consumption [4].

Effect of dipping treatment with different concentration of EDTA (75, 125, 200 ppm) and KMS (500 ppm) was studied on weight loss and other quality parameters in strain **SSI 4035** of *A. bisporus* (**Table 3**). It was noticed that the weight loss (0.94 g) in fruiting bodies preserved under ambient temperature conditions in different concentrations of EDTA and KMS was more during 2–3 days of storage period with no much effect on whiteness (**Table 4**) and texture (**Table 5**) but gill opening (**Table 6**) and condensation (**Table 7**) was greatly influenced. On the contrary, weight loss in fruiting bodies (0.53 g) was less when preserved in different concentrations of EDTA and KMS for 6–10 days of storage under refrigerated conditions with no effect on whiteness and texture but gill opening and condensation was considerably influenced. Pretreatments of mushrooms with (chlorinated) water, dipping in citric acid, sodium chloride, or KMS, blanching in hot water, blanching followed with soaking in whey and curd fermentation, or steam blanching followed by sulfiting and citric acid were used by several workers before drying to stabilize color, flavor enhancement and texture retention [20–23]. The earlier findings also corroborates the present results in stabilizing the whiteness and retention of texture of fruiting bodies of *A. bisporus*.

Treatment	Weight loss %			Texture (days)			Whiteness (days)			Gill opening (days)		
	First year	Second year	Average	First year	Second year	Average	First year	Second year	Average	First year	Second year	Average
Ambient temperature												
EDTA 75 ppm	0.15	2.15	1.15	3	4	3.5	2	6	4	3	5	4
EDTA 125 ppm	0.27	1.78	1.02	3	4	3.5	2	6	4	4	5	4.5
EDTA200 ppm	0.42	1.85	1.13	3	4	3.5	3	6	4.5	4	5	4.5
EDTA 500 ppm	0.33	0.97	0.56	3	4	3.5	2	5	3.5	3	5	4
Refrigerated												
EDTA 75 ppm	0.48	0.82	0.65	10	14	12	11	16	13.5	11	15	13
EDTA 125 ppm	3.41	0.68	2.04	10	14	12	9	16	12.5	10	15	12.5
EDTA200 ppm	0.53	0.55	0.54	11	13	12	11	16	13.5	11	15	13
EDTA 500 ppm	0.50	0.70	0.6	8	13	11	9	11	10	9	15	12

Table 1.
Effect of dipping treatment on weight loss, whiteness, texture and gill opening of Agaricus boisporus *cv. S 649.*

Treatment	Weight loss %			Texture (days)			Gill opening (days)			Whiteness (days)		
	First year	Second year	Average	First year	Second year	Average	First year	Second year	Average	First year	Second year	Average
Ambient temperature												
EDTA 75 ppm	0.76	0.88	0.82	4	4	4	4	5	4.5	3	4	3.5
EDTA 125 ppm	0.84	1.24	1.04	4	4	4	4	5	4.5	3	4	3.5
EDTA200 ppm	0.93	1.23	1.08	4	4	4	4	5	4.5	3	4	3.5
KMS 500 ppm	0.77	1.32	1.04	4	4	4	4	5	4.5	3	4	3.5
Refrigerated												
EDTA 75 ppm	0.93	0.60	0.76	16	14	15	16	13	14.5	16	16	16
EDTA 125 ppm	0.88	0.72	0.8	16	14	15	16	7	11.5	11	16	16
EDTA200 ppm	0.53	0.58	0.55	16	14	15	16	8	12	12	16	14
KMS 500 ppm	0.49	0.48	0.48	16	12	14	16	8	12	16	10	13

Table 2.
Effect of dipping treatment on weight loss, whiteness, texture and gill opening of Agaricus boisporus *cv. NCS 100.*

Treatment	Weight of fruiting bodies (g)																			
	Initial	1DA	2DA	3DA	4DA	5DA	6DA	7DA	8DA	9DA	10DA	11DA	12DA	13DA	14DA	15DA	16DA	17DA	18DA	Over all reduction
									Ambient temperature											
EDTA 75 ppm	40.56	40.29 (0.27)	40.02 (0.27)	39.72 (0.30)	—	—	—	—	—	—	—	—	—	—	—	—	—	—	—	0.84
EDTA 125 ppm	38.71	38.38 (0.33)	38.01 (0.37)	37.67 (0.34)	—	—	—	—	—	—	—	—	—	—	—	—	—	—	—	1.04
EDTA 200 ppm	47.06	46.78 (0.28)	46.51 (0.27)	46.25 (0.24)	—	—	—	—	—	—	—	—	—	—	—	—	—	—	—	0.81
KMS 500 ppm	39.37	38.99 (0.38)	38.65 (0.34)	38.30 (0.35)	—	—	—	—	—	—	—	—	—	—	—	—	—	—	—	1.07
									Refrigerated temperature											
EDTA 75 ppm	25.09	25.03 (0.06)	24.99 (0.04)	24.95 (0.04)	24.90 (0.05)	24.86 (0.04)	24.81 (0.05)	24.75 (0.06)	24.72 (0.03)	24.68 (0.04)	26.63 (0.05)	24.57 (0.06)	24.53 (0.04)	24.47 (0.06)	24.40 (0.07)	24.35 (0.05)	24.29 (0.06)	24.20 (0.09)	24.15 (0.05)	0.94
EDTA 125 ppm	33.73	33.67 (0.06)	33.62 (0.05)	33.58 (0.04)	33.54 (0.04)	33.48 (0.06)	33.44 (0.04)	33.38 (0.06)	33.31 (0.07)	33.29 (0.02)	33.23 (0.06)	33.17 (0.06)	33.12 (0.05)	33.07 (0.04)	33.05 (0.03)	33.00 (0.05)	32.90 (0.10)	32.83 (0.07)	32.74 (0.09)	0.99
EDTA 200 ppm	26.47	26.39 (0.08)	26.34 (0.05)	26.30 (0.04)	26.24 (0.06)	26.20 (0.04)	26.14 (0.06)	26.09 (0.05)	26.02 (0.07)	25.99 (0.03)	25.92 (0.07)	25.84 (0.08)	25.79 (0.03)	25.73 (0.06)	25.68 (0.05)	25.64 (0.04)	25.50 (0.14)	25.44 (0.06)	25.35 (0.09)	1.12
KMS 500 ppm	32.25	32.14 (0.11)	32.07 (0.07)	32.01 (0.06)	31.94 (0.07)	31.89 (0.05)	31.83 (0.06)	31.76 (0.07)	31.72 (0.04)	31.67 (0.05)	31.61 (0.06)	31.54 (0.07)	31.48 (0.06)	31.42 (0.06)	31.37 (0.05)	31.32 (0.05)	31.20 (0.12)	31.11 (0.09)	31.01 (0.10)	1.14

Table 3.
Effect of dipping treatments on weight loss of Agaricus bisporus *fruit bodies SSI 4035.*

DOI: http://dx.doi.org/10.5772/intechopen.101168

Treatment	Whiteness of fruiting body*																			
	Initial	1DA	2DA	3DA	4DA	5DA	6DA	7DA	8DA	9DA	10DA	11DA	12DA	13DA	14DA	15DA	16DA	17DA	18DA	19DA
Ambient temperature																				
EDTA 75 ppm	1	1	1	3	—	—	—	—	—	—	—	—	—	—	—	—	—	—	—	—
EDTA 125 ppm	1	1	1	3	—	—	—	—	—	—	—	—	—	—	—	—	—	—	—	—
EDTA 200 ppm	1	1	1	2	3	—	—	—	—	—	—	—	—	—	—	—	—	—	—	—
KMS 500 ppm	1	1	1	3	—	—	—	—	—	—	—	—	—	—	—	—	—	—	—	—
Refrigerated temperature																				
EDTA 75 ppm	1	1	1	1	1	1	1	1	1	1	1	1	1	1	1	1	2	2	3	3
EDTA 125 ppm	1	1	1	1	1	1	1	1	1	1	1	1	1	1	2	2	2	2	2	3
EDTA 200 ppm	1	1	1	1	1	1	1	1	1	1	1	1	1	1	1	2	2	2	2	3
KMS 500 ppm	1	1	1	1	1	1	1	1	1	1	1	1	1	1	1	1	2	2	2	3

1—absolute white, 2—slight dull (cream), 3—light brown and Dark brown.

Table 4.
Effect of dipping treatments on whiteness of Agaricus bisporus *fruit bodies SSI 4035.*

Treatment	Gill opening after days of storage*																			
	Initial	1DA	2DA	3DA	4DA	5DA	6DA	7DA	8DA	9DA	10DA	11DA	12DA	13DA	14DA	15DA	16DA	17DA	18DA	19DA
Ambient temperature																				
EDTA 75 ppm	1	2	5	5	—	—	—	—	—	—	—	—	—	—	—	—	—	—	—	—
EDTA 125 ppm	1	5	5	5	—	—	—	—	—	—	—	—	—	—	—	—	—	—	—	—
EDTA 200 ppm	1	1	5	5	—	—	—	—	—	—	—	—	—	—	—	—	—	—	—	—
KMS 500 ppm	1	3	5	5	—	—	—	—	—	—	—	—	—	—	—	—	—	—	—	—
Refrigerated temperature																				
EDTA 75 ppm	1	1	1	1	1	1	1	2	3	3	3	3	3	3	3	3	4	4	4	5
EDTA 125 ppm	1	1	1	1	1	1	1	2	3	4	4	5	—	—	—	—	—	—	—	—
EDTA 200 ppm	1	1	1	1	1	2	2	5	5	—	—	—	—	—	—	—	—	—	—	—
KMS 500 ppm	1	1	1	1	1	1	2	2	3	3	3	3	3	4	4	4	4	5	—	—

1—without opening, 2—1/4 opening, 3—1/2 opening, 4—3/4 opening and 5—full open.

Table 5.
Effect of dipping treatments on gill opening of Agaricus bisporus *fruit bodies SSI 4035.*

DOI: http://dx.doi.org/10.5772/intechopen.101168

Treatment	Texture after days of storage																			
	Initial	1DA	2DA	3DA	4DA	5DA	6DA	7DA	8DA	9DA	10DA	11DA	12DA	13DA	14DA	15DA	16DA	17DA	18DA	19 DA
Ambient temperature																				
EDTA 75 ppm	1	1	1	3	—	—	—	—	—	—	—	—	—	—	—	—	—	—	—	—
EDTA 125 ppm	1	1	1	3	—	—	—	—	—	—	—	—	—	—	—	—	—	—	—	—
EDTA 200 ppm	1	1	1	2	—	—	—	—	—	—	—	—	—	—	—	—	—	—	—	—
KMS 500 ppm	1	1	1	3	—	—	—	—	—	—	—	—	—	—	—	—	—	—	—	—
Refrigerated temperature																				
EDTA 75 ppm	1	1	1	1	1	1	1	1	1	1	1	2	2	2	2	2	2	2	2	2
EDTA 125 ppm	1	1	1	1	1	1	1	1	1	1	1	1	1	1	1	1	1	1	1	1
EDTA 200 ppm	1	1	1	1	1	1	1	1	1	1	1	2	2	2	2	2	2	2	2	2
KMS 500 ppm	1	1	1	1	1	1	1	1	1	1	2	2	2	2	2	2	2	2	2	2

1—smooth, 2—rough, 3—not acceptable.

Table 6.
Effect of dipping treatments on texture of Agaricus bisporus *fruit bodies SSI 4035.*

Treatment	Condensation																			
	Initial	1DA	2DA	3DA	4DA	5DA	6DA	7DA	8DA	9DA	10DA	11DA	12DA	13DA	14DA	15DA	16DA	17DA	18DA	19DA
Ambient temperature																				
EDTA 75 ppm	1	2	2	2	—	—	—	—	—	—	—	—	—	—	—	—	—	—	—	—
EDTA 125 ppm	1	2	2	2	—	—	—	—	—	—	—	—	—	—	—	—	—	—	—	—
EDTA 200 ppm	1	2	3	3	—	—	—	—	—	—	—	—	—	—	—	—	—	—	—	—
KMS 500 ppm	1	2	3	3	—	—	—	—	—	—	—	—	—	—	—	—	—	—	—	—
Refrigerated temperature																				
EDTA 75 ppm	1	1	2	2	2	2	2	2	2	2	2	2	2	2	2	2	2	2	2	2
EDTA 125 ppm	1	1	2	2	2	2	2	2	2	2	2	3	3	3	3	3	3	3	3	3
EDTA 200 ppm	1	1	2	3	3	3	3	3	3	3	3	3	3	3	3	3	3	3	3	3
KMS 500 ppm	1	1	2	2	2	2	2	2	2	2	2	2	2	2	2	2	2	2	2	2

1—1/4 of pp. bag, 2—1/2 of pp. bag, 3—3/4 of pp. bag and 4—full of pp. bag.

Table 7.
Effect of dipping treatments on condensation of Agaricus bisporus *fruit bodies SSI 4035.*

Effect of dipping treatment with different concentration of EDTA (75–200 ppm) and KMS (500 ppm) was studied on quality parameters of *Agaricus bisporus cv.* Pant 31 under ambient as well as refrigerated conditions (**Tables 8–11**). It was noticed that the fruiting bodies remained well up to 14–16 days under refrigerated conditions with least effect on whiteness, texture, condensation and opening of the gills of fruiting body whereas, the fruiting bodies remained well up to only 03 days under ambient temperature conditions with least effect on whiteness (**Table 8**), texture (**Table 9**), condensation (**Table 10**) and opening of the gills of fruiting body (**Table 11**). It was surprising to note that the weight loss in fruiting bodies preserved in different concentrations of EDTA (75–200 ppm) and KMS (500 ppm) for 14–16 days of storage were almost same under refrigerated conditions (0.54 g) and 03 days of storage under ambient temperature conditions (0.51 g). Washing of mushroom in different types of washing treatments (0.5%

Treatment	Whiteness of fruiting bodies days after storage																				
	Initial	1	2	3	4	5	6	7	8	9	10	11	12	13	14	15	16	17	18	19	20
Ambient temperature																					
EDTA 75 ppm	1	1	1	1	1	2	4	—	—	—	—	—	—	—	—	—	—	—	—	—	—
EDTA 125 ppm	1	1	1	1	1	1	2	3	4	—	—	—	—	—	—	—	—	—	—	—	—
EDTA 200 ppm	1	1	1	1	1	1	2	2	2	3	4	—	—	—	—	—	—	—	—	—	—
KMS 500 ppm	1	1	1	2	3	3	4	—	—	—	—	—	—	—	—	—	—	—	—	—	—
Refrigerated																					
EDTA 75 ppm	1	1	1	1	1	1	1	1	1	1	1	1	1	1	1	1	1	1	1	1	2
EDTA 125 ppm	1	1	1	1	1	1	1	1	1	1	1	1	1	1	1	1	1	1	1	1	1
EDTA 200 ppm	1	1	1	1	1	1	1	1	2	2	2	3	3	3	4	—	—	—	—	—	—
KMS 500 ppm	1	1	1	1	1	1	1	1	1	1	1	1	1	1	1	1	1	1	1	1	2

Table 8.
Effect of dipping treatments on whiteness of Agaricus bisporus *fruit bodies CV. Pant 31.*

Treatment	Texture after days of storage																				
	Initial	1	2	3	4	5	6	7	8	9	10	11	12	13	14	15	16	17	18	19	20
Ambient temperature																					
EDTA 75 ppm	1	1	1	1	1	1	1	—	—	—	—	—	—	—	—	—	—	—	—	—	—
EDTA 125 ppm	1	1	1	1	1	1	1	1	—	—	—	—	—	—	—	—	—	—	—	—	—
EDTA 200 ppm	1	1	1	1	1	1	1	1	1	1	—	—	—	—	—	—	—	—	—	—	—
KMS 500 ppm	1	1	2	2	2	2	2	—	—	—	—	—	—	—	—	—	—	—	—	—	—
Refrigerated temperature																					
EDTA 75 ppm	1	1	1	1	1	1	1	1	1	1	1	1	1	1	1	1	1	1	1	1	1
EDTA 125 ppm	1	1	1	1	1	1	1	1	1	1	1	1	1	1	1	1	1	1	1	1	1
EDTA 200 ppm	1	1	1	1	1	1	1	1	1	2	2	3	4	4	4	5	—	—	—	—	—
KMS 500 ppm	1	1	1	1	1	1	1	1	1	1	1	1	1	1	1	1	1	1	1	1	1

Table 9.
Effect of dipping treatments on texture of Agaricus bisporus *fruit bodies CV. Pant 31.*

Treatment	Condensation days after storage																				
	Initial	1	2	3	4	5	6	7	8	9	10	11	12	13	14	15	16	17	18	19	20
	Ambient temperature																				
EDTA 75 ppm	1	1	1	2	2	2	3	—	—	—	—	—	—	—	—	—	—	—	—	—	—
EDTA 125 ppm	1	1	1	2	2	3	3	3	—	—	—	—	—	—	—	—	—	—	—	—	—
EDTA 200 ppm	1	1	1	2	2	3	3	3	3	3	—	—	—	—	—	—	—	—	—	—	—
KMS 500 ppm	1	1	1	2	2	2	2	—	—	—	—	—	—	—	—	—	—	—	—	—	—
	Refrigerated temperature																				
EDTA 75 ppm	1	2	2	2	2	2	2	2	3	3	3	3	3	3	3	3	3	3	3	3	3
EDTA 125 ppm	1	2	2	2	2	2	3	3	3	3	3	3	3	3	3	3	3	3	3	3	3
EDTA 200 ppm	1	1	2	2	2	2	2	2	2	—	—	—	—	—	—	—	—	—	—	—	—
KMS 500 ppm	1	2	2	2	2	2	2	2	3	3	3	3	3	3	3	3	3	3	3	3	3

Table 10.
Effect of dipping treatments on condensation of Agaricus bisporus *fruit bodies CV. Pant 31.*

Treatment	Gill opening days after storage																				
	Initial	1	2	3	4	5	6	7	8	9	10	11	12	13	14	15	16	17	18	19	20
	Ambient temperature																				
EDTA 75 ppm	1	1	1	1	1	1	1	—	—	—	—	—	—	—	—	—	—	—	—	—	—
EDTA 125 ppm	1	1	1	1	2	2	3	5	—	—	—	—	—	—	—	—	—	—	—	—	—
EDTA 200 ppm	1	1	1	1	1	1	1	1	1	1	1	—	—	—	—	—	—	—	—	—	—
KMS 500 ppm	1	1	1	1	1	1	1	—	—	—	—	—	—	—	—	—	—	—	—	—	—
	Refrigerated temperature																				
EDTA 75 ppm	1	1	1	1	1	1	1	1	1	1	1	1	1	1	1	1	1	1	1	1	2
EDTA 125 ppm	1	1	1	1	1	1	1	1	1	1	1	2	3	4	5	—	—	—	—	—	—
EDTA 200 ppm	1	1	1	1	1	1	1	1	1	1	1	1	1	1	1	1	1	1	1	1	1
KMS 500 ppm	1	1	1	1	1	1	1	1	1	1	1	1	1	1	1	1	1	1	1	1	1

Table 11.
Effect of dipping treatments on gill opening of Agaricus bisporus *fruit bodies CV. Pant 31.*

calcium chloride and 0.5% citric acid, combination wash with 50 ppm chlorine dioxide, 0.1% sodium erythrobate and 0.05% calcium chloride, and 0.5–1.5% potassium metabisulphite) was also examined by others with fairly good results [4, 24, 25] to reduce postharvest spoilage of mushrooms.

2.2 Effect of packaging thickness on quality parameters of button mushroom

2.2.1 Effect of packaging thickness on quality of cv. S 649 of A. bisporus

Effect of packaging thickness on quality parameters of two button mushroom cv. S 649 and cv. NCS 100 was studied under ambient temperature and refrigerated conditions at Mushroom Research Laboratory of the Department of Plant Pathology, IGKV, Raipur. Transparent polyethylene bags of three thickness (75,100 and

S. no.	Treatments	At ambient temperature							
		Weight (% weight loss)		Veil opening			Color		
		Initial	After 24 h	Initial	After 24 h	After 48 h	Initial	After 24 h	After 48 h
1	75 Gauge	34.63	34.50 (0.37)	+	++	+++	+	+++	++++
2	100 Gauge	39.16	39.08 (0.20)	+	++	+++	+	+++	++++
3	125 Gauge	45.00	44.61 (0.86)	+	++	+++	+	+++	++++
4	100 PE (control)	38.56	38.29 (0.70)	+	++	+++	+	+++	++++

For color −*+: absolute white, ++: dull white, +++: dull yellow, ++++: not acceptable.*
For veil opening – *+: intact fruit body, ++: veil opening up to 25%, +++: veil opening up to 50%, ++++: veil opening up to 100%.*

Table 12.
Packaging for button mushroom cv. S 649 (washed 0.05% KMS).

125 gauge) along with one control (100 PE) was used. The fruiting bodies of 25 ± 2 cm or available size fruit bodies of cv. S 649 of *A. bisporus* were first washed for 10 min in washing solution containing 0.05% KMS and stored at ambient temperature and refrigerated temperature conditions for different durations.

The effect of packing thickness on weight loss, veil opening and color of the fruit bodies of cv. S 649 of *A. bisporus* was studied at ambient temperature and the results are presented in **Table 12**. At Ambient temperature, the weight loss in fruiting bodies of strain cv. S 649 of *A. bisporus* varied from 0.20 to 0.86% in varying thickness of polyethylene bags. The weight loss was maximum (0.86%) in 125 gauge thickness while, it was minimum (0.2%) in 100 gauge of thickness after 24 h of storage. The color became dull yellow and veil opening was also noticed in all the treatments after 24 h. The maximum weight loss (98.08%) in the untreated oyster mushroom was also recorded at the 7th day of storage by Das et al. [4]. However, the lowest weight loss (33.62%) was observed in oyster mushrooms when it was wrapped in unperforated plastic bag. Similarly, quality parameters like protein content was found to be higher (28.98%) in oyster mushrooms wrapped with unperforated plastic bag followed by perforated plastic bag (25.00%) at the 5th day of storage.

At refrigerated temperature conditions, maximum weight loss (6.78%) in fruit bodies of strain cv. S 649 was recorded in 75 gauge pp. bags and minimum weight loss (2.72%) was noticed in 125 gauge after 5 days of storage (**Table 13**). In general, weight loss in fruiting bodies of *A. bisporus* was less under refrigerated conditions and more under ambient temperature conditions but the trend here was just reversed which is difficult to explain under same set of conditions. The fruit bodies of *A. bisporus* retained absolute white color up to 24 h in all the treatments. The color was then changed from absolute white to dull white after 48 h in all the treatments except 100 gauge pp. bags. Dull white color was consistently retained up to 5 days in all the treatments. Veil opening was not noticed in any of the treatments up to 24 h except control. Thereafter, veil opening was observed up to 25% in 75 and 125 gauge PP bags whereas 50% veil opening was observed in 100 gauge pp. bags and control up to 5 days of storage. The storage of oyster mushrooms in modified atmosphere packaging (MAP) was found by Das et al. [4] to be very effective in reducing moisture loss. It may be mainly due to storing perishables in MAP which regulates gaseous exchange, reduces weight loss, spoilage, and maintains quality of mushrooms during postharvest handling.

S. no.	Treatments	At refrigerated temperature																	
		Weight (% weight loss) after hours						Color after hours						Veil opening after hours					
		Initial	24	48	72	96	120	Initial	24	48	72	96	120	Initial	24	48	72	96	120
1	75 Gauge PP	34.50	33.26 (2.50)	33.06 (4.17)	32.80 (4.92)	32.63 (5.42)	32.16 (6.78)	+	+	++	++	++	++	+	+	++	++	++	++
2	100 Gauge PP	38.00	37.81 (0.50)	37.63 (0.97)	37.19 (2.13)	36.98 (2.92)	36.78 (3.21)	+	+	+	++	++	++	+	+	++	++	++	+++
3	125 Gauge PP	51.40	51.08 (0.62)	50.75 (1.26)	50.38 (1.98)	50.13 (2.47)	50.00 (2.72)	+	+	++	++	++	++	+	+	++	++	++	++
4	100 PE (control)	38.63	38.50 (0.33)	38.10 (1.37)	37.76 (2.52)	37.18 (3.75)	36.93 (4.40)	+	+	++	++	++	++	+	++	++	++	+++	+++

For color —+: absolute white, ++: dull white, +++: dull yellow, ++++: not acceptable.
For veil opening – +: intact fruit body, ++: veil opening up to 25%, +++: veil opening up to 50%, ++++: veil opening up to 100%.

Table 13.
*Packaging for button mushroom (*Agaricus bisporus *cv. S 649) (washed 0.05% KMS).*

Treatments	At room temperature								
	Weight (% weight loss)			Color			Veil Opening		
	Initial	After 24 h	After 48 h	Initial	After 24 h	After 48 h	Initial	After 24 h	After 48 h
75 Gauge	33.4	32.89 (1.52)	32.54 (2.57)	+	+	++++	+	+	++
100 Gauge	44.77	43.91 (1.92)	43.29 (3.30)	+	+	++++	+	+	++
125 Gauge	39.36	38.89 (1.19)	38.55 (2.05)	+	+	++++	+	+	+++
100 PE (control)	41.8	41.33 (1.12)	41.00 (1.91)	+	+	++++	+	+	+++

For color: +: absolute white, ++: dull white, +++: dull yellow, ++++: not acceptable.
For veil opening: +: intact fruit body, ++: veil opening up to 25%, +++: veil opening up to 50%, ++++: veil opening up to 100%.

Table 14.
Packaging for button mushroom cv. NCS 100 washed with KMS (0.05%).

2.2.2 Effect of packaging thickness on quality strain cv. NCS 100 of A. bisporus

Effect of packaging thickness on weight loss, color and veil opening of the fruiting bodies of strain cv. NCS 100 of *A. bisporus* was studied at ambient temperature and refrigerated conditions (**Table 14**). At room temperature, maximum weight loss (3.30%) was recorded in 100 gauge pp. bags while, it was minimum (1.91%) in control but the extent of losses in cv. NCS 100 was more compared to cv. S 649. The fruit bodies of *A. bisporus* retained absolute white color up to 24 h in all treatments. Thereafter, they were not acceptable. No veil opening was observed up to 24 h in all the treatments. However, 25% veil opening was noticed in 75 and 100 gauge and it was 50% in 125 gauge and control after 48 h. The quality parameters in cv. NCS 100 were better compared to the cv. S 649 except weight loss of the fruiting bodies.

At refrigerated temperature, maximum weight loss (1.39%) in cv. NCS 100 was recorded in 75 gauge pp. bags and minimum weight loss (1.05%) was noticed in control after 5 days of storage (**Table 15**) which was remarkably less compared to the fruit bodies of cv. S 649. The fruit bodies of cv. NCS 100 of *A. bisporus* retained absolute white color up to 4 days in 125 gauge pp. bags while 2 days in other treatments. Thereafter, the fruit bodies changed to dull white color in 125 gauge on 5th day and other treatments from 3 to 4 days. The color was further changed to dull yellow in other treatments on 5th day. Veil opening was not at all noticed in any of the treatments up to 5 days of storage. In all the quality parameters, the fruiting bodies of cv. NCS 100 was far better than cv. S 649. Thus, it can be very well said that the fruiting bodies of cv. NCS 100 can be very well preserved up to 5 days under refrigerated conditions with least influence on different quality parameters compared to cv. S 649. Longer shelf life of fruiting bodies of cv. S 649 might be due to slow respiration rate under refrigerated conditions noticed during present investigation.

2.3 Evaluation of different methods to enhance the shelf life of oyster mushroom

Drying is one of most broadly-practiced and oldest methods for preserving agricultural products to maintain the quality against decaying. It is done mainly in warmer areas such as the kitchen, near the stove or fireplace. These are used as heat sources and often drying is completed in the sun [26]. Three methods of oyster

Treatments	At refrigerated temperature																	
	Weight (% weight loss)						Color						Veil opening					
	Initial	After 24 h.	After 48 h.	After 72 h.	After 96 h.	After 120 h.	Initial	After 24 h.	After 48 h.	After 72 h.	After 96 h.	After 120 h.	Initial	After 24 h.	After 48 h.	After 72 h.	After 96 h.	After 120 h
75 Gauge	37.35	37.26 (0.24)	37.19 (0.42)	37.07 (0.74)	37.01 (0.91)	36.83 (1.39)	+	+	+	++	++	+++	+	+	+	+	+	+
100 Gauge	36.76	36.69 (0.19)	36.63 (0.35)	36.48 (0.76)	36.40 (01.97)	36.29 (1.27)	+	+	+	++	++	+++	+	+	+	+	+	+
125 Gauge	37.8	37.73 (0.18)	37.67 (0.34)	37.60 (0.52)	37.44 (0.84)	37.3 (1.32)	+	+	+	+	+	++	+	+	-	+	+	+
100 PE (control)	36.91	36.86 (0.13)	36.78 (0..35)	36.72 (0.51)	36.63 (0.75)	36.52 (1.05)	+	+	+	++	++	+++	+	+	+	+	+	+

For color: +: absolute white, ++: dull white, +++: dull yellow, ++++: not acceptable.
For veil opening : +: intact fruit body, ++: veil opening up to 25%, +++: veil opening up to 50%, ++++: veil opening up to 100%.

Table 15.
*Packaging for button mushroom (*Agaricus bisporus *cv. NCS 100) washed with KMS (0.05%).*

mushroom preservation/drying were evaluated under farmers conditions. The oyster mushroom (200 g) were first blanched at 75°C for 2 min in water. Thereafter, it was kept in two steeping solutions of 5% salt, 0.2% citric acid, 0.15% potassium meta bisulfite (KMS). In the second treatment, the same quantity of mushroom was dried using mechanical dryer at 45–50°C for 8 h. In third treatment, the same quantity of oyster mushroom was blanched and sun dried. Blanching with the use of hot water or steam is an important treatment applied after washing to inhibit tissue browning by inactivation of polyphenol oxidase and production of off flavors. It also removes trapped air and decreases weight losses to induce mushroom shrinkage [27]. These were then observed for color, texture and overall acceptability. In another experiment, freshly harvested fruiting body (500 g) of oyster mushroom (*Pleurotus florida*) was wrapped in muslin cloth and blanched in chemical solution (0.2% salt and 0.1% citric acid) at 75°C for 2 min. The fruiting bodies so obtained were sun dried and also dried in cabinet dryer at 60°C for 6.30 h. The freshly harvested fruiting body were also immerged in plain water and dried by sun as well as by cabinet dryer and the results are presented in **Table 16**.

Of three treatments assessed, the fresh oyster mushroom steeped in solutions of above chemicals with or without blanching were of good quality for the period of 105 days without much loss in color, texture and acceptability. Mushroom dried in a mechanical dryer at 45–50°C for 8 h. With blanching though changed in to blackish color but the loss in weight, brittleness and quality of the fruiting body remained least influenced followed by without blanching which were of excellent quality even up to 3 months of storage. The sun dried product when dipped in plain water and kept in pp. bags after 55 days at ambient temperature lost maximum weight of the fruiting body (3.12%) than any other treatment. The fruiting body became soft and developed brown color. Thus, it can be said that the oyster mushroom steeped in solutions of above chemicals and mechanically dried can be very well preserved for the period of 105 days without much influence on quality parameters. The increase in the drying temperature though helped to accelerate the drying rate but, high temperatures (70°C) are not generally recommended because, it causes browning in the samples and deteriorates the quality which are important from customer's viewpoint [28]. In another experiment conducted at AICMIP, Raipur center on different methods of drying of oyster mushroom exhibited drying by cabinet dryer

S. no.	Treatments	Weight gain /loss		Change in Color	Brittle ness*	Quality parameters (3 months)**
		Initial	After 3 months (% wt. loss)			
1.	Sun drying with blanching (0.2% salt +0.1% CA for 2 min)	29.63	29.0 (2.12)	Blackish	+	++
2.	Sun drying without blanching (0.2% salt +0.1% CA for 2 min)	37.07	36.03 (2.80)	Light Brown	++	+
3.	Cabinet drying with blanching	33.11	32.75 (1.08)	Blackish	+	+
4.	Cabinet drying without blanching	32.07	31.44 (1.96)	Yellowish	+	+
5.	Dipping in plain water followed by sun drying (Control)	36.50	35.36 (3.12)	Light Brown	++	+

NB. There was no rottage and insect attack within the storage period of 3 months.
**+ Brittle, ++ Soft,*
***Quality parameters: + Pleasant flavor ++ Off flavor.*

Table 16.
*Studies on methods of drying of oyster mushroom (*Pleurotus florida*) on weight loss and other quality parameters.*

with and without blanching to be excellent as there was minimum reduction in weight, color was retained, fruit body was brittle and pleasant flavor was noticed within the period of 3 months of storage (**Table 16**). Other methods of drying were not that much effective as the per cent reduction in weight was more, fruit body started turning yellowish in color, became soft and developed off flavor. The results obtained on the preservation method had significant effects on the nutrient and mineral compositions of the mushroom samples [29]. In contrast to present investigation, the lowest weight values were obtained from the sundried mushroom samples while the highest value was obtained from the fresh samples by Jonathan et al. [30] which is difficult to be explained.

2.4 Post harvest treatments and its effect on shelf life of *Pleurotus flabellatus*

In order to study the quality and storage, the sporophores of *P. flabellatus* with and without blanching were steeped in solutions of different chemicals (**Tables 17–20**). The fresh sporophores (150 g) were blanched at 98°C for 2 min. Using double layer of

Treatment		****Color of sporophores in 1–5 scale at different time intervals (days)**								
Number	**Treatment details***	**1**	**5**	**25**	**50**	**75**	**100**	**125**	**150**	**175**
T-1	5% salt, 0.2% C.A., 0.1% KMS (WOB)	1	4	5						
T-2	5% salt, 0.2% C.A., 0.1% KMS (WB)	1	3	5						
T-3	2% salt, 1% sugar, 0.3% C.A., 0.1% KMS (WB)	1	3	5						
T-4	2% salt, 1% sugar, 0.3% C.A., 0.1% KMS (WOB)	1	3	5						
T-5	2.5% salt, 0.1% A.A., 0.2% C.A.,0.1% S.B. 0.1% KMS (WB)	1	1	1	1	2	2	2	3	3
T-6	2.5% salt, 0.1% A.A., 0.2% C.A., 0.1% S.B. 0.1% KMS (WOB)	1	3	5						
T-7	0.2% A.A., 0.2% C.A., 0.2% KMS (WB)	1	1	1	1	2	2	2	3	3
T-8	0.2% A.A., 0.2% C.A., 0.2% KMS (WOB)	1	3	5						
T-9	0.5% C.A. (WB)	1	4	5						
T-10	Simple boiled water (WB)	1	4	5						
T-11	0.1% A.A., 0.2% P.A., 0.1% KMS (WB)	1	1	1	1	1	1	2	3	
T-12	5% salt, 0.2% C.A., 0.1% KMS (WB)	1	1	1	1	1	1	2	2	3
T-13	0.1% A.A., 0.3% C.A., 0.1% KMS, 1% ASA (WB)	1	2	3	4					
T-14	1% salt, 0.1% A.A., 0.1% C.A., 0.05% S.B. 0.05% KMS (WB)	1	1	1	1	1	1	2	2	3
T-15	0.1% A.A., 0.1% C.A., 0.1% KMS (WB)	1	1	1	1	1	1	2	3	
T-16	1% Salt, 0.1% C.A., 0.05% KMS (WB)	1	3	4						
T-17	0.1% KMS, 0.2%, A.A. (WB)	1	2	4						
T-18	0. 3% A.A. (W.B.)	2	4	5						
T-19	0.2% KMS (WB)	2	4	5						

**SB—sodium benzoate, ASA—ascorbic acid, C.A.—citric acid, KMS—potassium metabisulphide.*
***Scale white—1, like white—2, slight dull—3, A.A.—acitic acid, P.A.—propionic acid, W.B.—with blanch, WOB —without blanch, light brown—4, dark brown—5.*

Table 17.
Effect of different steeping solution on color and storage of the sporophores of Pleurotus flabellatus.

Treatment		**Texture of sporophoresin 1–7 scale at different time intervals (days)								
Number	**Treatment details***	**1**	**5**	**25**	**50**	**75**	**100**	**125**	**150**	**175**
T-1	5% salt, 0.2% C.A., 0.1% KMS (WOB)	1	7							
T-2	5% salt, 0.2% C.A., 0.1% KMS (WB)	1	5							
T-3	2% salt, 1% sugar, 0.3% C.A., 0.1% KMS (WB)	1	5							
T-4	2% salt, 1% sugar, 0.3% C.A., 0.1% KMS (WOB)	1	5							
T-5	2.5% salt, 0.1% A.A., 0.2% C.A.,0.1% S.B. 0.1% KMS (WB)	1	1	2	2	2	2	2	3	3
T-6	2.5% salt, 0.1% A.A., 0.2% C.A., 0.1% S.B. 0.1% KMS (WOB)	1	4							
T-7	0.2% A.A., 0.2% C.A., 0.2% KMS (WB)	1	1	2	2	2	2	2	3	3
T-8	0.2% A.A., 0.2% C.A., 0.2% KMS (WOB)	1	4							
T-9	0.5% C.A. (WB)	1	7							
T-10	Simple boiled water (WB)	1	7							
T-11	0.1% A.A., 0.2% P.A., 0.1% KMS (WB)	1	2	2	2	2	2	3	3	
T-12	5% Salt, 0.2% C.A., 0.1% KMS (WB)	1	2	2	2	2	2	2	3	3
T-13	0.1% A.A., 0.3% C.A., 0.1% KMS, 1% ASA (WB)	1	2	2	4					
T-14	1% salt, 0.1% A.A., 0.1% C.A., 0.05% S.B. 0.05% KMS (WB)	1	2	2	2	2	2	2	3	3
T-15	0.1% A.A., 0.1% C.A., 0.1% KMS (WB)	1	2	2	2	2	2	3	3	
T-16	1% salt, 0.1% C.A., 0.05% KMS (WB)	1	5							
T-17	0.1% KMS, 0.2%, A.A. (WB)	1	2	4						
T-18	0. 3% A.A. (W.B.)	1	7							
T-19	0.2% KMS (WB)	1	7							

*SB—sodium benzoate, ASA—ascorbic acid, C.A.—citric acid, KMS—potassium metabisulphide.
**Scale fresh—1, like fresh—2, less sogy—3, more sogy—4, A.A.—acetic acid, P.A—propionic acid, W.B—with blanch, WOB—without blanch, coarse—5, rotting—6, leathery—7.

Table 18.
Effect of different steeping solution on texture and storage of the sporophores of Pleurotus flabellatus.

muslin cloth. Thereafter, the sporophores were transferred in the steeping solutions prepared from various chemicals and their concentrations forming sum of 19 treatments. The steeping solution of 500 ml was taken in a plastic container of 1 L capacity and lid was screwed. These containers were then stored at room temperature and observations on colors, texture, appearance and overall acceptability (in days) were recorded following different scales at different time intervals.

2.4.1 Effect of steeping solution on color

The observations of the study indicated that the treatments, T12 and T14 can retain good color (2) of the sporophores of *P. flabellatus* till 150 days (**Table 17**). Thereafter, it became slight dull (3) but it was too acceptable up to 175 days. It was followed by T5, T7, T11 and T15 which could equally preserve the sporophores till 125 days and up to the acceptable color by 150 days. In remaining treatments, the acceptable color of the sporophores could not be maintained even up to 5 days. The sporophores kept under these treatments started quick deterioration. The steeping

Treatment		**Appearance of sporophores in 1–6 scale at different time intervals (days)								
Number	**Treatment details***	**1**	**5**	**25**	**50**	**75**	**100**	**125**	**150**	**175**
T-1	5% salt, 0.2% C.A., 0.1% KMS (WOB)	2	6							
T-2	5% salt, 0.2% C.A., 0.1% KMS (WB)	2	4							
T-3	2% salt, 1% sugar, 0.3% C.A., 0.1% KMS (WB)	2	4							
T-4	2% salt, 1% sugar, 0.3% C.A., 0.1% KMS (WOB)	2	4							
T-5	2.5% salt, 0.1% A.A., 0.2% C.A., 0.1% S.B. 0.1% KMS (WB)	1	2	2	2	2	2	3	3	3
T-6	2.5% salt, 0.1% A.A., 0.2% C.A., 0.1% S.B. 0.1% KMS (WOB)	2	4							
T-7	0.2% A.A., 0.2% C.A., 0.2% KMS (WB)	1	2	2	2	2	2	3	3	3
T-8	0.2% A.A., 0.2% C.A., 0.2% KMS (WOB)	2	4							
T-9	0.5% C.A. (WB)	2	6							
T-10	Simple boiled water (WB)	2	6							
T-11	0.1% A.A., 0.2% P.A., 0.1% KMS (WB)	1	2	2	2	2	2	3	3	
T-12	5% salt, 0.2% C.A., 0.1% KMS (WB)	1	2	2	2	2	2	3	3	3
T-13	0.1% A.A., 0.3% C.A., 0.1% KMS, 1% ASA (WB)	1	2	2	6					
T-14	1% salt, 0.1% A.A., 0.1% C.A., 0.05% S.B. 0.05% KMS (WB)	1	2	2	2	2	2	3	3	3
T-15	0.1% A.A., 0.1% C.A., 0.1% KMS (WB)	1	2	2	2	2	2	3	3	
T-16	1% Salt, 0.1% C.A., 0.05% KMS (WB)	2	4							
T-17	0.1% KMS, 0.2%, A.A. (WB)	2	2	6						
T-18	0. 3% A.A. (W.B.)	2	6							
T-19	0.2% KMS (WB)	2	6							

**SB—sodium benzoate, ASA—ascorbic acid, C.A.—citric acid, KMS—potassium metabisulphide,*
***Scale fresh—1, very good—2, good—3, fair—4, A.A.—acitic acid, P.A—propionic acid, W.B—with blanch, WOB —without blanch, slight fermented smell—5, unacceptable—6.*

Table 19.
Effect of different steeping solution on appearance and storage of the sporophores of Pleurotus flabellatus.

solutions became turgid, less transparent and profuse growth of the fungal contaminants occurred on the top of the steeping solution. It was observed that the sporophores kept after blanching could better retain the color in comparison to without blanched sporophores. It was also noticed that the lower concentration of the steeping solutions (T15) worked equally well compared to that of higher concentration (T7).

2.4.2 Effect of steeping solutions on texture

The texture of the sporophores preserved in steeping solutions of T12 and T14 was almost fresh (2) up to 125 days and was acceptable (3) up to 175 days (**Table 18**). In T11 and T15, the sporophores were like fresh (2) till 100 days and up to the acceptable period of 150 days. In rest of the treatments, the sporophores preserved in steeping solutions of different chemicals showed more sogginess (4), rotted (6) leathery (7) and became unacceptable for consumption within 5 to 25 days. In T2, T9, T10, T18, T19, the sporophores of *P. flabellatus* exhibited fast

Treatments	Treatment details*	Storage period of quality sporophores (days)
T-1	5% salt, 0.2% C.A., 0.1% KMS (WOB)	3–4
T-2	5% salt, 0.2% C.A., 0.1% KMS (WB)	4–5
T-3	2% salt, 1% sugar, 0.3% C.A., 0.1% KMS (WB)	4–5
T-4	2% salt, 1% sugar, 0.3% C.A., 0.1% KMS (WOB)	4–5
T-5	2.5% salt, 0.1% A.A., 0.2% C.A.,0.1% S.B.0.1% KMS (WB)	165–175
T-6	2.5% salt, 0.1% A.A., 0.2% C.A., 0.1% S.B. 0.1% KMS (WOB)	5–6
T-7	0.2% A.A., 0.2% C.A., 0.2% KMS (WB)	165–175
T-8	0.2% A.A., 0.2% C.A., 0.2% KMS (WOB)	5–6
T-9	0.5% C.A. (WB)	2–3
T-10	Simple boiled water (WB)	2
T-11	0.1% A.A., 0.2%P.A., 0.1% KMS (WB)	150–155
T-12	5% Salt, 0.2% C.A., 0.1% KMS (WB)	165–175
T-13	0.1% A.A., 0.3% C.A., 0.1% KMS, 1% ASA (WB)	45–48
T-14	1% salt, 0.1% A.A., 0.1% C.A., 0.05% S.B. 0.05% KMS (WB)	165–175
T-15	0.1% A.A., 0.1% C.A., 0.1% KMS (WB)	150–155
T-16	1% Salt, 0.1% C.A., 0.05% KMS (WB)	5–6
T-17	0.1% KMS, 0.2%, A.A. (WB)	20–21
T-18	0. 3% A.A. (W.B.)	2
T-19	0.2% KMS (WB)	2

**SB—sodium benzoate, ASA—ascorbic acid, C.A.—citric acid, KMS—potassium metabisulphide. A.A.—acetic acid, P.A—propionic acid, W.B—with blanch, WOB—without blanch.*

Table 20.
Effect of steeping solution on storage and quality of sporophores of Pleurotus flabellatus.

deterioration starting from 2 to 3 days. The part of the sporophores get rotted and dissolved in steeping solutions.

2.4.3 Effect of steeping solution on appearance of sporophorus

Sporophores of *P. flabellatus* preserved in steeping solutions of T5, T7, T11, T12, T14 and T15 appeared well (2) up to 100 days and were good up to 150–175 days of storage period (**Table 19**). In other treatments, the sporophores kept with and without blanching and steeped in solutions of different chemicals seemed to be fair (4) and became unacceptable (6) in appearance within 5–25 days of storage. The appearance of the sporophores after blanching and steeping in lower concentrations of the chemical was extremely good.

2.4.4 Effect of steeping solutions on quality and shelf life during storage

It is evident from **Table 20** that T5, T7, T12 and T14 preserved the mushroom (*P. flabellatus)* up to 165–175 days (5.5 months) followed by T-11 and T-15

(150–155 days) without any adverse effect in color, texture, appearance and overall acceptability.

Mushroom preservation was reported to be of great importance due to off-season household consumption among geographically spread groups of the society [31]. The shelf life of *P. flabellatus* fruit bodies were almost doubled and that too with lower concentration of chemical solutions and without much change in color, texture appearance and acceptability (5.5 months). It may possibly be due to more toughness, leatheryness and bigger size of *P. flabellatus* fruit bodies which might have sustained the effect of chemicals for longer period of time as against the fruit body of *P. sajor-caju* which remained comparatively smaller, thinner and less leathery due to which it could have preserved well only up to 3 months [5]. The present results are also in contradiction with the work of earlier scientists who reported higher shelf life (24 days) of paddy straw mushroom with higher concentration of chemicals and lower shelf life (13 days) with lower concentration of chemicals [32]. In T 13, the fruit bodies of *P. flabellatus* were well preserved for 45–48 days which is in agreement with the findings of Sethi et al. [8] and Pruthi et al. [9]. Blanching of oyster mushroom in present investigation was found to be the most important process in avoiding the microbial and other deteriorations which was also reported by Absule et al. [5] and Bano and Singh [33] with respect to oyster and white button mushrooms respectively.

2.5 Mushroom processing and value addition

The rural women were very well aware with the naturally growing edible mushrooms which they do collect during monsoon season. In the villages, it is a common practice to dehydrate naturally growing mushroom in a local Bhatti and use it in a season when it is not available. With this background, the interested women were selected for training programme on Mushroom Processing Technology. Before selection, they were interviewed for necessity of such training programs. They emphasized that the fresh mushroom is not being sold many a times and it becomes difficult for them to preserve it. Oyster mushroom which were cultivated by them was sun dried as per our directives.

Seventy-six women from Tarra, Dondekhurd and Matia villages were selected for the training on Mushroom Processing Technology. For establishment of processing units at Tarra and Dondekhurd, solar dryers and mechanical tray dryers for drying of mushrooms were procured which are used by the women for drying of edible mushrooms. Oyster mushroom after dehydration was grind into powder, sieved and added @ 10–25% in preparation of various mushroom processed products. Mushroom powder was added in preparation of some of the local products like murku, bijoura, chakli etc. by Thakur [14, 15]. The training programs on Mushroom Processing Technology was mainly imparted to the women in preparation of Mushroom Mung Papad, Mushroom Urd Papad, Steamed Rice Mushroom Papad, Mushroom Badi, Mushroom Soup, Mushroom Bijora, Mushroom Murku, Mushroom Chakli, Mushroom Biscuits, Mushroom Pickles. The women of all the villages very much liked the training on this aspect. The mushroom dishes prepared by the women were highly appreciated during Peer Review Team. They were willing to sale the processed mushroom products in the local market. Mushroom pickle, mushroom badi and mushroom papad became very much popular among the women and they see very good future of these products in Chhattisgarh Market.

An attempt was made to establish the linkages with the mushroom entrepreneurs working in different parts of Chhattisgarh. The persons involved in this business were called at the villages, samples were given to them and they assured the disposal of their fresh and processed mushroom products. It has also been tried to display their products in Business Counseling Centers to be established by Swa Shakti Project in the rural areas.

2.6 Conclusion

It was found that the button mushrooms and their varieties when steeped in solutions of different chemicals (EDTA and KMS) and packaging materials (thickness) were able to sufficiently enhance the shelf life under both ambient and refrigerated conditions. Similarly, drying methods particularly cabinet drying at 45–60°C with blanching or without blanching did play an important role in extension of shelf life of oyster mushroom for 3 months under ambient conditions without much influence on quality parameters. Oyster mushroom when preserved in different chemical solutions of lower concentrations, it was found to prolong the shelf life of fresh oyster mushroom up to 6 months without much influencing the quality parameters. Similarly, the shelf life of dried oyster mushroom was also enhanced for 105 days even preserving at ambient temperature conditions. Processing of mushroom and their value addition in preparing various local or traditional dishes incorporating mushrooms had a great role to play in extending the shelf life and minimizing the existing problem of under nourishment and mal nourishment prevailing in several states of India particularly in weaker section of the society. Thus, there is an urgent need to promote mushroom processed products like royal oyster capsule, mushroom fortified wheat flour, mushroom based mung/urd nuggets, mushroom based mug/urd papad, mushroom rice papad, mushroom instant soup, mushroom pickle, mushroom powder, mushroom soup etc. so as to popularize them among the mankind and promote marketing of these products.

Author details

Mahesh Prasad Thakur[1]*, Harvinder K. Singh[2] and Chandra Shekhar Shukla[2]

1 Faculty of Agriculture, Indira Gandhi Krishi Vishwavidyalaya, Raipur, Chhattisgarh, India

2 Department to Plant Pathology, College of Agriculture, Indira Gandhi Krishi Vishwavidyalaya, Raipur Chhattisgarh, India

*Address all correspondence to: mp_thakur@yahoo.com

References

[1] Diamantopoulou PA, Philippoussis AN. Cultivated mushrooms: Preservation and processing. In: Hui YH, Evranuz EO, editors. Handbook of Vegetable Preservation and Processing. Boca Raton: CRC Press; 2015. pp. 495-525. DOI: 10.1201/b19252-26. Available from: https://www.researchgate.net/publication/297715290

[2] Mahajan PV, Oliveira FAR, Macedo I. Effect of temperature and humidity on the transpiration rate of the whole mushrooms. Journal of Food Engineering. 2008;**84**:281-288

[3] Burton KS, Frost CE, Nichols R. A combination plastic film system for controlling post harvest mushroom quality. Mushroom News. 1989;**37**(7):6-10

[4] Das PK, Hassan MK, Akhther N. Efficacy of washing and postharvest treatments on shelf life and quality of oyster mushroom. Progressive Agriculture. 2010;**21**(1&2):21-29

[5] Adsule PG, Girija V, Dan A, Tewari RP. A note of simple preservation of oyster mushroom (*Pleurotus sajor-caju*). Indian Journal of Mushrooms. 1981;7(1&2):2-5

[6] Gogoi P, Baruah. Studies on steeping preservation of fresh edible mushroom with particular reference to *Pleurotus* and *Volvariella volvacea*. In: Indian Mushroom Conference 1997. Souvenir and Abstract. Solan (H.P.): National Research Centre for Mushroom; 1997. p. 74 (Abs.)

[7] Singh A, Kesherwani GP, Gupta OP. Dehydration and steeping preservation of paddy straw mushroom (*Volvariella volvacea)*. Mushroom Research. 1996;**5**: 39-42

[8] Sethi V, Bahl N, Bhagwan J. Steeping preservation of mushroom (*Agaricus bisporus)*. In: Symp. on Impact of Pollution in and from Food Industry and its Management. Mysore: CFTRI; 1989. p. 50 (Abstr.)

[9] Pruthi JS, Manon JK, Raina BL, Teotia MS. Improvement in whiteness and extension of shelf life of fresh and processed mushrooms (*Agaricus bisporus* and *Volvariella volvacea)*. Indian Food Packer. 1984;**38**(2):55-63

[10] Gormley TR, O' Riordain F. Quality evaluation of fresh and processed oyster Pleurotus Ostreatus. Lebensumittel-Wissenschaft und Technologie. 1976;**9**:75

[11] Martinez-Soto G, Paredes LO, Ocana CR, Bautista JM. Oyster mushroom (*Pleurotus ostreatus)* quality as affected by modified atmosphere packaging. Micologia Neotropial Aplicada. 1998;**11**:53-67

[12] Mehta KB, Jandaik CL. Storage and dehydration studies of fresh fruit bodies of dhingri mushroom—*Pleurotus sapidus*. Indian Journal of Mushroom. 1989;**15**:17-2

[13] Rajarathnam S, Bano Z, Patwardhan MV. Post harvest physiology and storage of the white oyster musrhoom (*Pleurotus. flabellatus)*. Journal of Food Science and Technology. 1983;**18**(2):153-162

[14] Thakur MP. Advances in post-harvest technology and value additions of edible mushrooms. Indian Phytopathology. 2018;**71**(3):303-315

[15] Thakur MP. Advances in mushroom production: key to food, nutritional and employment security: A review. Indian Phytopathology. 2020;**73**(3):337-395. DOI: 10.1007/s42360-020-00244-9

[16] Mehta BK, Jain SK, Sharma GP, Doshi A, Jain HK. Cultivation of button mushroom and its processing: A

techno-economic feasibility. International Journal of Advanced Biotechnology and Research. 2011;**2**: 201-207

[17] Burton KS, Twyning RV. Extending mushroom storage life by combining modified atmosphere packaging and cooling. Acta Horticulturae. 1989;**258**: 565-571

[18] Wakchaure GC. Postharvest handling of fresh mushrooms. In: Singh M, Vijay B, Kamal S, Wakchaure GC, editors. Mushrooms: Cultivation, Marketing and Consumption. Solan: Directorate of Mushroom Research, Indian Council of Agricultural Research (ICAR); 2011. pp. 197-206

[19] Gormley TR. Chill storage of mushrooms. Journal of the Science of Food and Agriculture. 1975;**26**(4): 401-411

[20] Kotwaliwale N, Bakane P, Verma A. Changes in textural and optical properties of oyster mushroom during hot air drying. Journal of Food Engineering. 2007;**78**:1207-1211

[21] Martínez-Soto G, Ocana-Camacho R, Paredes-Lopez O. Effect of pretreatment and drying on the quality of oyster mushrooms (*Pleurotus ostreatus*). Drying Technology. 2001;**19** (3–4):661-672

[22] Singh SK, Narain M, Kumbhar BK. Effect of drying air temperatures and standard pretreatments on the quality of fluidized bed dried button mushroom (*Agaricus bisporus*). Indian Food Packer. 2001;**55**(5):82-86

[23] Walde SG, Velu V, Jyothirmayi T, Math RG. Effects of pretreatments and drying methods on dehydration of mushroom. Journal of Food Engineering. 2006;**74**(1):108-115

[24] Jayathunge L, Illeperuma C. Extension of postharvest life of oyster mushroom under ambient conditions by modified atmosphere packaging. Journal of Tropical Agricultural Research. 2001; **13**:78-89

[25] Jayathunge L, Illeperuma C. Extension of postharvest life of oyster mushroom by modified atmosphere packaging technique. Journal of Food Science. 2005;**70**(9):E573-E578

[26] Moreno FA. Un recurso alimentario de los grupos originarios y mestizos de México: Los hongos silvestres. Anales de Antropología. 2014;**48**:241-272

[27] Wu CM, Wu JLP, Chen CC, Chou CC. Flavor recovery from mushroom blanching water. In: Charalambous G, Inglett G, editors. The Quality of Foods and Beverages: Chemistry and Technology. Vol. 1. New York: Academic Press; 1981. pp. 133-145

[28] Demiray E. Effect of drying temperature on color and desorption characteristics of oyster mushroom. Food Science and Technology. 2020; **40**(1):187-193

[29] Zhu D, Guo R, Wenxiang L, Jingya S, Cheng F. Improved postharvest preservation effects of Pholiota nameko mushroom by sodium alginate–based edible composite coating. Food and Bioprocess Technology. 2019;**12**:587-598. DOI: 10.1007/s11947-019-2235-5

[30] Jonathan GS, Omotayo OO, Baysah GI, Asemoloye MD, Aina DA. Effects of some preservation methods on the nutrient and mineral compositions of three selected edible mushrooms. Journal of Microbial & Biochemical Technology. 2018;**4**: 106-111. DOI: 10.4172/1948-5948.1000402

[31] Ruan-Soto F, Ordaz-Velázquez M, García-Santiago W, Pérez-Ovando EC. Traditional processing and preservation of wild edible mushrooms in Mexico.

Annals of Food Processing and Preservation. 2017;**2**(1):1013

[32] Ramaswamy K, Kandaswamy TK. Possible cause for the quick deterioration of quality of paddy straw mushroom in storage. Indian Mushroom Science. 1978;**1**:329-335

[33] Bano Z, Singh NS. Steeping preservation of a edible mushroom (*Agaricus bisporus*). Journal of Food Science and Technology. 1972;**9**(1):13-15